# The Speaking Animal

# The Speaking Animal

## *Ethics, Language and the Human-Animal Divide*

Alison Suen

ROWMAN & LITTLEFIELD
INTERNATIONAL
London • New York

Published by Rowman & Littlefield International, Ltd.
Unit A, Whitacre Mews, 26-34 Stannary Street, London SE11 4AB
www.rowmaninternational.com

Rowman & Littlefield International, Ltd. is an affiliate of Rowman & Littlefield
4501 Forbes Boulevard, Suite 200, Lanham, Maryland 20706, USA
With additional offices in Boulder, New York, Toronto (Canada), and London (UK)
www.rowman.com



**British Library Cataloguing in Publication Information Available**
A catalogue record for this book is available from the British Library

ISBN: HB 978-1-78348-511-6
ISBN: PB 978-1-78348-512-3

**Library of Congress Cataloging-in-Publication Data**

Suen, Alison, author.
The speaking animal : ethics, language and the human-animal divide / Alison Suen.
pages cm
Includes bibliographical references and index.
ISBN 978-1-78348-511-6 (cloth : alk. paper) -- ISBN 978-1-78348-512-3 (pbk.) -- ISBN 978-1-78348-513-0 (electronic)
1. Animal rights--Philosophy. 2. Speciesism--Philosophy. 3. Speech--Philosophy. 4. Animal communication--Philosophy. I. Title.
HV4708.S84 2015
179'.3--dc23
2015026628

∞™ The paper used in this publication meets the minimum requirements of American National Standard for Information Sciences Permanence of Paper for Printed Library Materials, ANSI/NISO Z39.48-1992.

Printed in the United States of America

To my parents—both in Hong Kong and in Iowa

# Contents

# Acknowledgements

This book argues that kinship is indispensable to our capacity to speak—it is by developing our capacity to relate that we become speaking animals. As such, I would like to acknowledge my parents, who taught me my first words; Trevor M. Bibler, who not only read the entire manuscript, but also makes funny jokes; and Linguini, who relentlessly challenges the cat-human language divide with her purrs and meows. I thank my family for being the bedrock on which I develop my academic voice.

I am grateful to all who have offered encouragement, advice, and criticism on this book. I owe special gratitude to Kelly Oliver and Rebecca Tuvel for their support and friendship. I am grateful to my editor, Sarah Campbell, whose comments and encouragement made the process of writing and publishing a pleasure. Thanks also to Sinéad Murphy for her patience and attentive editorial assistance. Thanks to the following individuals who had read and offered invaluable comments on my drafts at various stages: Geoff Adelsberg, Ellen Armour, Ryan Brand, Garrett Bredeson, Adam Burgos, Lisa Guenther, Leonard Lawlor, Juliana Lewis, José Medina, and Namrata Mitra. I would like to thank my colleagues at Iona College for their collegiality, especially Alex Eodice, Michael Jordan, and Chris Perricone for their mentorship. I would also like to thank the following individuals who, in different ways, have helped me along the way: Susanne Baltes, Elizabeth Edenberg, Jeffrey Tlumak, Yuki, and Sunny.

A different version of chapter 2 was published as “From Animal Father to Animal Mother: A Freudian Account of Animal Maternal Ethics.” *philoSOPHIA* 3 (2): 121–37.

# Introduction

Is there a connection between our capacity to speak and our responsibilities towards others? The slogan that we need to "give voice to the voiceless" suggests that this is the case: insofar as we can speak, we have an obligation to speak for those who cannot speak for themselves. This slogan is particularly poignant in the context of animal advocacy—not only do animals lack a *political* voice, they seem to lack the capacity to speak altogether. So what are our responsibilities to animals? What do we owe other animals by virtue of our status as "the animal with *logos*"?

In the history of philosophy, humans are said to be unique amongst animals because we are the only animals with *logos*. But if language is *the* capacity that separates us from other animals, then the dictum that we need to give voice to the voiceless animal becomes a reminder of our uniqueness: the very effort to speak for the animal highlights the fact that we are distinct from the animal for which we speak. In other words, animal advocacy is predicated on a human-animal language divide: every time we attempt to speak for the animal, we reinforce the human-animal language divide.

Is there a way to give voice to the voiceless animal without deepening the divide or presupposing human exceptionalism? And is there a way to articulate our identity as the speaking animal without pitting the human *against* the animal? After all, it seems incongruent to try to transform our relationship with the animal while simultaneously highlighting our purported superiority, especially when this purported superiority is often employed to justify animal exploitation. In *The Speaking Animal*, I consider our ethical obligations, *as* speaking animals, to other animals. I argue that we need to be sensitive to the violence of animal advocacy without disavowing our responsibility to speak, and to speak well. To this end, we need to reconsider what it means to be a speaking animal, and what makes language possible in the first place.

Going against the philosophical current that ties language to reason primarily, in *The Speaking Animal* I argue that language capacity is first and foremost a relational capacity—our capacity to speak hinges on our capacity to develop kinship with others, sometimes even kinship across species. In the end, our identity as speaking animal is constructed not by pitting ourselves against the animal, but rather by recognizing our kinship with them. I will trace pertinent moments in twentieth-century philosophy wherein linguistic capacity and relational capacity are intimately linked. While the philosophers I examine come from different traditions, with radically different philosophical agendas and commitments, their accounts of language allow us to uncover the relational possibilities found at the core of language.

Before I move onto my chapter-by-chapter synopsis, I would like to manage some expectations. Although I am critical of the linguistic divide that continues to reassert itself in animal advocacy, my ambition is not to challenge the purported exclusivity of our language capacity. In other words, I will not argue that animals also speak, nor will I argue that they exhibit rudimentary or proto-language such that our linguistic difference is merely a difference in degree—this is simply not the debate I engage here.

What I argue in this book is that we need to *reinterpret* the human-animal linguistic divide. Traditionally, the linguistic divide works as follows: we self-identify as speaking animals by setting ourselves in opposition to dumb beasts. As such, the validation of our identity presupposes a dichotomous human-animal relationship. In the following pages, I will argue that we need to reconceptualize language, as well as our identity as linguistic beings. Instead of subscribing to the idea that language is a divisive capacity that bifurcates humans from animals, we need an account of language that is grounded in our social, relational capacity. It is by reinterpreting, rather than denying, the uniqueness of our identity as the speaking animal that we can properly take on our responsibility to speak for those who lack a voice.

While the recent proliferation of discourse in animal ethics/animal philosophy has certainly enriched my research, it also makes the question of *Why another book on animals?* more pressing. Accordingly, in chapter 1 I motivate my overall project by situating it within animal rights literature. It examines how the linguistic divide has shaped the current discourse in animal ethics. Since Aristotle declared humans to be the only "animal with *logos*," the linguistic divide between humans and animals has found various incarnations in the history of philosophy. Against this backdrop, philosophers in the field of animal ethics have dutifully addressed the language divide. Following the lead of utilitarian philosophers such as Jeremy Bentham and Peter Singer, they maintain that one's capacity to speak is irrelevant to one's moral status. While I agree with them that linguistic capacity should not determine moral status, I argue that the efficacy of the animal rights movement continues to depend on one's capacity to speak. In fact, the power of rights is very much informed by our power to speak.

Drawing from feminist writings on the problem of "speaking for the other," chapter 1 addresses the problem of "speaking for the animal." The works of Linda Alcoff and Catharine MacKinnon are especially illuminating. Some critics of the rights discourse are concerned with the "violence" of speaking for the animal; specifically, our impulse to speak on behalf of the animal can be a form of paternalism that may very well obliterate the voice of the animal. These critics suggest that a non-anthropocentric animal ethics requires that we appreciate the voice—as well as the silence—of animals. While I am sympathetic to their concerns, I argue that a non-anthropocentric ethics is a fantasy at best and a hindrance to animal welfare at worst. So instead of rejecting the power of speech, I motivate my readers to interrogate what it means to have a voice, and to consider an account of language that prioritizes kinship and imagination, rather than reason.

While chapter 1 identifies some internal problems and tensions within animal advocacy, chapter 2 problematizes the limited vision of the current discourse. My main target here is the abolitionist approach in animal ethics, in particular its narrow focus on what we ought *not* do to the animal. Although this approach forms a great alliance with the legal discourse that pervades the literature, it remains reactionary insofar as it merely *negates* the status quo. I argue that we need an animal ethics (and an account of animality) that goes beyond the rhetoric of abolition and prohibition; we need an animal ethics that acknowledges the regulatory power of love, creativity, and imagination. I turn to an unlikely ally—Sigmund Freud—to explore a positive way to articulate our relationship with animals.

To set up my argument, I investigate Freud's study of infantile zoophobias. According to Freud, in nearly all cases of infantile animal phobia, the feared animal functions as a father figure. The feared animal takes on the prohibitive role as a father-substitute. The substitutability of the animal and the father is crucial for Freud, as it anchors his theory regarding the familial, social, and religious structure of patriarchal societies. In light of this standard animal-father substitution, Freud's biography of Leonardo da Vinci stands out as a provocative exception. In this psychoanalytic biography, Freud examines da Vinci's relationship with a vulture—only here the vulture is an androgynous creature that serves as a *mother*-substitute. I argue that Freud's interpretation of da Vinci opens up a new way to understand our relationships with animals—a way that not even Freud himself anticipated. In short, I analyse the significance of this deviant case of animal obsession in Freud's corpus and its ramifications for reconceiving the human-animal relationship.

Whereas the first two chapters offer critical analyses of the current animal rights discourse, chapter 3 begins the constructive work of *The Speaking Animal*. To develop what I call a "relational account of language," I turn to

Martin Heidegger, who, unlike many philosophers, is critical of the link between language and reason throughout the history of philosophy. There is already a significant amount of literature on Heidegger's views of animal and animality. Many have critiqued Heidegger's notorious claim that animals are "poor in world" in *The Fundamental Concepts of Metaphysics*, the lecture course in which he most extensively articulates his concept of the "animal." While the "poverty thesis" is typically examined in relation to Heidegger's anthropocentrism, my focus in this chapter is different. My argument revolves around Heidegger's discussion of the word "symbol" in this lecture course, a discussion that has largely been overlooked by commentators.

Following Aristotle, Heidegger isolates "symbol" as the key ingredient of human discourse. Heidegger describes "symbol" as the joining together of two things. The primary image he uses to illustrate this "joining together" is two halves of a friendship token. As my chapter shows, for Heidegger there is a parallel between the animal's supposed lack of language and the animal's supposed lack of kinship with humans. By denying animals language, Heidegger is also denying them kinship with us. Building on Heidegger's etymological analysis of "symbol," I argue that the condition for discourse—*logos*—depends on kinship.

Whereas in chapter 3 I trace the root of language to kinship, in chapter 4 I use Ludwig Wittgenstein's notion of "family resemblance" to make sense of our kinship with, as well as our estrangement from, animals. Specifically, I address how we can acknowledge our kinship with animals without doing away with our differences or denying our uniqueness. I begin this chapter by exploring a topic that remains under-theorized in animal ethics: animal identity. The question of identity is intimately tied to the question of language, specifically the question of *how* we should speak for animals. The first part of this chapter introduces and substantiates the problems of identity politics and standpoint theory within animal ethics. Most notably, feminists have critiqued the strategy of the "like-us" model in animal ethics; this logic of identification is problematic insofar as it can inadvertently reinstate the human exceptionalism that animal advocates vehemently reject. My chapter is situated at the crossroads of identification and counter-identification: must an animal ethics be grounded in what we share in common with the animal? Or is an animal ethics of difference possible and even necessary? My position is that this is a false dilemma: a Wittgensteinian account of family resemblance resists both a sharp divide and a linear continuity between human and animal.

Following José Medina's "familial view of identity," I turn to Wittgenstein's notion of family resemblance. I show that the metaphor of family offers a fruitful way of articulating the diversity found within the category "animal." It allows us to speak of the concept of "animal" without homogenizing animals: just as the concept of "language" is constituted by a variety of language games, the concept of "animal" is constituted by a variety of subgroups (families) that overlap and crisscross. Insofar as identity and difference are dependent on each other, we cannot employ the concept of "animal" without recognizing its inherent multiplicity, and we cannot speak of our kinship with animals without acknowledging that we are also strangers.

At different moments in *The Speaking Animal* I shall argue that we need to own our unique capacity to speak. If we cannot avoid speaking for the animal, then we must seek a good and responsible way to speak for them—or be minimally sensitive to any injurious and irresponsible ways of speaking for them. The concluding chapter is a critical analysis of animal advocacy in popular culture and political discourse, with specific attention paid to racism and xenophobia. I examine the ways in which racist and xenophobic discourse appropriates the rhetoric of animal rights. I analyse *The Cove* (2009), an Oscar-winning documentary that has mobilized the recent outcry over dolphin hunting in Japan. I identify two typical tactics of dehumanization in *The Cove* that unwittingly racialize and marginalize the Japanese dolphin hunters: (1) the use of the language divide to demarcate hero and villain, and (2) the use of the cruelty-animality link to demonstrate the less-than-human status of animal abusers. But as I shall argue, as long as humanness is linked to humanity and cruelty to animality, the human-animal dichotomy remains intact in animal advocacy. Indeed, animal advocates are often complicit in the anthropocentrism that they seek to challenge precisely because they continue to employ speciesist language even as they speak for animals. Until we interrogate and challenge the link between beastliness and cruelty, we risk dehumanizing and marginalizing racial minorities in the name of animal advocacy.

In sum, *The Speaking Animal* is an effort to reconsider what it means to be an animal with *logos*—not just in terms of who we are, but also how this identity informs our difference from, and responsibility to, the animal other. We need not disavow the uniqueness of our capacity to speak: uniqueness does not necessarily translate into human superiority, nor does it presuppose an abyss between humans and animals. By articulating a relational account of language, I hope to show that we become most human, and most willing to speak, when we acknowledge our kinship with the animal other.

# Chapter 1

# Giving Animals a Hearing

## *Rights Discourse and Animal Representation in Animal Ethics*

In *The Politics*, Aristotle posits humans as the only animals endowed with speech, with *logos* (1253a). Aristotle's definition of the human proliferates throughout the history of Western philosophy, where the human is often defined against other animals by virtue of this capacity to speak. This linguistic divide has been translated into an ontological hierarchy whereby the human is privileged over the animal—an ontological hierarchy that, many animal advocates contend, informs our treatment of animals.

Against this philosophical backdrop, philosophers concerned with animal ethics have dutifully addressed this linguistic divide. Many of them have recourse to Jeremy Bentham's famous line, "the question is not, Can they *reason*? nor, Can they *talk*? but, Can they *suffer*?"[1] For Bentham, sentience, not language, entitles one to ethical considerations.[2] Bentham's view has been popularized by Peter Singer's utilitarian defence of animal liberation[3] and is further reiterated by other animal rights philosophers. Richard Sorabji, for example, insists that the linguistic capacity of animals is "a point of the highest scientific interest, but of absolutely no moral relevance whatsoever" (1993, 216).[4] Gary Francione echoes Sorabji's sentiment: "the short answer to the question posed by Nagel's essay is *who cares what it is like to be a bat?* As long as the bat is sentient, then whatever other characteristics the bat has or does not have are *irrelevant* for the moral purpose of whether we should treat the bat exclusively as our resource" (2000, 120). In an article a few years later, Francione responds to the claim that humans are uniquely linguistic with a simple question: "so what?" (2009, 8). Within the animal rights discourse, the question of linguistic difference is often deemed irrelevant to the issue of moral status.[5]

Other philosophers have responded differently by challenging the boundary of the linguistic divide.[6] Their effort is twofold. On the one hand, they employ empirical studies of animal communication as evidence against a rigid linguistic divide. They ask, if dolphins can learn American Sign Language,[7] if Koko the gorilla can make up new words, lie to her trainer, and even learn to rhyme,[8] how can we still insist on defining humans in terms of our capacity to speak? How can we maintain the exceptionality of humans by appealing to language? On the other hand, philosophers also point to "marginal cases"[9] where humans do not develop the capacity to speak. According to these arguments from marginal cases, if we are unwilling to exploit those humans who can't speak, then linguistic difference clearly cannot be the justification for our exploitation of animals. Once again, the linguistic divide is shown to be an impertinent criterion for determining moral status.

While the question of linguistic capacity is ritualistically addressed in the literature, it is usually broached in the context of demonstrating its moral irrelevance. (It is as if philosophers want to address the linguistic question in order to get it out of the way.) While I agree that linguistic capacity should not govern one's moral status,[10] we can fruitfully look at the linguistic question from a different angle by considering the *social-political* significance of being a speaking subject.

In what follows, I examine the connection between "having a right" and "having a say" in animal ethics. I identify various ways in which the animal rights discourse remains logocentric by privileging language and those who can speak.[11] I will consider specifically the rhetoric of "giving a voice to the voiceless" in animal advocacy, wherein the advocates and the animals are dichotomously split between the speaking and the voiceless. Given that the task of an animal advocate involves speaking for the animal, it is important to consider the power dynamic between the advocate and the animal.[12]

After showing the privileged status of language in the literature and identifying some of the problems of speaking for animals, I will examine two proposed solutions to these problems: (1) letting animals speak for themselves, and (2) recognizing the special power of silence. Although these two solutions are meant to curb the "violence" of speaking for animals, I argue that they are problematic in their own right. I end this chapter by demonstrating the necessity of speaking well for animals. In order to speak well for animals, we need to reconceptualize language and reconsider what it means to be a speaking animal.

## WHAT IS WRONG WITH ANIMAL RIGHTS?

Anyone who has taken a quick glance at a university library collection on animal ethics immediately notices that the notion of animal *rights* dominates the literature. Mainstream discussions of animal ethics are often narrowed to the question of whether animals have rights and, if so, what sort of rights they have.[13] Peter Singer's *Animal Liberation* has had a tremendous influence on the animal rights movement, even though he claims that he is arguing for the liberation, not the rights, of animals.[14]

Despite its dominance, the animal rights discourse has been scrutinized by critics from various traditions. An early critique comes from Cora Diamond, who persuasively argues that the question of whether animals have rights is a misguided question; it is, in fact, "a totally wrong way of beginning the discussion" (1978, 467). Diamond argues that at stake is not whether animals have rights but whether animals are considered food (1978, 468). After all, the reason we do not eat people is not that they have the right not to be eaten (although they *do* have such a right); rather, we do not eat people because we do not see people as food.

Whereas Diamond argues that the question of animal "rights" is the wrong question, some critics are worried that the very concept of rights reinscribes oppression. A particularly prominent critical voice comes from feminists, in particular feminists in the care ethics tradition. Their criticisms of the rights discourse range from the problematic lineage of the concept of "rights" to the prioritization of reason over emotion.[15] For many care ethics feminists, the animal rights discourse continues to valorize a masculine philosophical tradition that is counterproductive, perhaps even antithetical, to the liberation of the oppressed—be it women or animals.

In addition to the "masculinist orientation" of the rights discourse (Kheel 2004, 336), critics are also worried that the *extension* of rights to animals is just another expression of human exceptionalism. Animal rights philosophy remains "speciesist" and "anthropocentric" as long as the inclusion in the moral community is dependent on whether one possesses certain features that we, as *humans*, consider important (Wolfe 2013, 13). Literary scholar Frances Bartkowski asks if the declaration of apes' rights would "entail new and more intimate ways of enslaving their capacities for reciprocal attachment" (2008, 86). In "Of Mice and Men," Catharine MacKinnon reminds us that "seeking animal rights on a 'like-us' model of sameness may be misconceived" because we "[miss] animals on their own terms" (2004, 264). For these critics, as long as the concept of animal rights is understood as an extension of human rights to a selected group of animals that are sufficiently *like humans*, there is a constant worry that animal rights may become a token of humanism or a vessel for human exceptionalism.

There is a further concern that the language of rights is predicated on the human-animal divide. For example, animal rights debates often fall back on a line-drawing strategy in order to demarcate creatures with rights from those without rights (Oliver 2009, 29). However, can the rights discourse truly overcome the human-animal divide if it continues to rest on the logic of exclusion and binary opposition? (Oliver 2009, 29). W. J. T. Mitchell articulates the paradox of animal rights as follows: "The very notion of 'animal rights,' to begin with, seems impossible insofar as it is modeled on human rights, because the very idea of *human* rights . . . is predicated on the difference between humans and animals" (2003, ix).

However, not all critics of the rights discourse find a rights-based moral framework problematic. Instead of jettisoning the language of rights altogether, some seek to *revamp* the animal rights framework. Most notably, in *Zoopolis* (2011) Sue Donaldson and Will Kymlicka argue that animal rights theory (ART) as it is currently conceived fails to effect systematic change in the institution of animal exploitation. However, it is not because ART continues to subscribe to a masculinist or anthropocentric moral framework, either by prioritizing reason over emotion, or by promoting human exceptionalism unwittingly. Rather, according to Donaldson and Kymlicka, ART fails because it has not sufficiently connected itself to the human rights model. The extension of rights to animals in the current framework is primarily confined to negative rights, such as the right *not* to be killed, *not* to be tortured, and so on. But just as humans have both negative and positive rights, animals also need both. The positive rights of animals are rarely entertained because, for many ART proponents, the best way to relate to animals is to leave them alone. In light of this halfhearted attempt to extend rights to animals in traditional ART, Donaldson and Kymlicka seek a moral framework that "connects the treatment of animals more directly to fundamental principles of liberal-democratic justice and human rights" (2011, 3). So rather than moving away from the human rights model, Donaldson and Kymlicka argue that ART needs to be more committed to the human rights model.[16]

While Donaldson and Kymlicka offer a reinterpretation of the animal rights framework that means to reinvigorate the animal rights debate, my ambition here is not to revamp or fine-tune the language of rights in animal ethics. As noted in the introduction, one of my goals in writing this book is to demonstrate the importance of the linguistic divide in animal ethics, despite philosophers' repeated attempts to downplay its relevance. Accordingly, I diagnose a version of human exceptionalism through the lens of the linguistic divide. I argue that the prioritization of rights is intertwined with the prioritization of language. The primacy of rights cannot be separated from the primacy of language *and* the privileging of linguistic beings. However, if animal ethics is no more than the extension of rights (whether it is negative or positive), and the very notion of rights presupposes the linguistic hierarchy, then we risk subjugating the animal even in our attempt to liberate them. The very means with which we try to liberate animals from our dominion can itself be an expression of our dominion.

## HAVING A RIGHT AND HAVING A SAY

In *The Alchemy of Race and Rights*, Patricia Williams responds to the critique of rights from Critical Legal Studies (CLS) scholars. She defends the rights discourse by speaking of the empowering effect of rights. For her, "one's sense of empowerment defines one's relation to the law" (1991, 148). While those who are already empowered by rights can afford to move beyond the rights discourse, those who have been deprived of rights still find the rights language desirable: "'Rights' feels new in the mouths of most black people. It is still deliciously empowering to say" (1991, 164). I am particularly interested in the *empowering* effect of rights. What does it mean to be empowered? Can we speak of empowerment for the animal?

According to the OED, to "empower" means "To invest with legal or formal power or *authority*; to authorize or license to do something or for a purpose" (Empower 2015). I emphasize "authority" here because of the etymological connection between authority and authorship[17]—having power means having the power to author.[18] Empowerment connects rights to the ability to author one's own speech, to speak for oneself. As Williams puts it, an expansion of rights "[gives] voice to those people or things that . . . historically have had no voice" (1991, 160). Rights give us a *say*, and without rights, we are as powerless as dumb animals. It is noteworthy that in her defence of rights, Williams compares her great-great-grandmother Sophie (who was a slave) to a wild fox. Whether a slave or a fox, "rights over them never filtered down to them; rights to their persons were never vested in them" (Williams 1991, 156). In light of the connection between rights and speech, how should we understand the empowering effects of *animal* rights? Does it mean letting animals speak for themselves? Must animals write for their rights? I will return to these questions later.

In addition to Williams's defence of rights, Jacques Derrida's reading of *Leviathan* in *The Beast and the Sovereign*, volume 1, also highlights the connection between rights and speech. In these posthumously published seminars, Derrida complicates the relationship among law, rights, and speech. Derrida's reading of *Leviathan* is characteristically rich, so for the purpose of this chapter I will focus on Derrida's reading of the following passage from *Leviathan*:

> To make Covenant with bruit Beasts, is impossible; because not understanding our speech, they understand not, nor *accept* of any translation of Right; nor can translate any Right to another: and without *mutuall acceptation*, there is no Covenant. (2009, 55; Derrida's emphasis)

Here we once again see how rights and speech intertwine. In Hobbes's view, animals do not enter into the covenant because they do not have language; and lacking language, they cannot accept any "translation of right."[19] As Derrida highlights in this passage, a key element of language is this ability to "accept"—the ability to respond and reciprocate (2009, 55). Without this responsivity, animals can neither "accept" rights nor commit themselves to a covenant. Importantly, Derrida observes that animals are not the only beings excluded from the Hobbesian covenant—God is also excluded from this covenant of men. More strikingly, God is excluded *for the same reason* as animals are excluded—neither God nor animals respond and reciprocate (2009, 55–57).

By drawing a parallel between the silence of God and the silence of the beast, Derrida complicate the speech-silence hierarchy. Silence (or unresponsiveness) is not always the mark of the dumb beast. More importantly, if God has a "right" to be silent, if God's sovereignty is expressed in this asymmetry of responsiveness, then why do we insist that the irresponsiveness of the animal is a defect, a lack, and even a justification for their exploitation? How does silence get bifurcated into the silence of the privileged and the silence of the oppressed?[20] If we perceive God's silent treatment as an exercise of power, can we also perceive animals' silence as powerful and meaningful in its own right? As we will see, silence can be an active form of resistance that conjures its own power.

Derrida also remarks that insofar as God and animals are excluded from the covenant, they both occupy a space external to the law. He even compares their "being-outside-the-law" status to that of the criminals (2009, 17). Whether "ignorant of right" or "having the right to suspend right," neither the animals nor God are subjects of—hence subjected to—the law (2009, 32). Thus far, the animal rights movement has responded dutifully to this "outside-the-law" status of animals. For animal rights advocates, the answer to the suffering of animals lies in undoing their "outlaw" status by including them in our "covenant," our law. While Hobbes would object to the inclusion of animals in the covenant, we see that the expansion of rights to animals reinforces, rather than challenges, the linguistic divide that defines the Hobbesian covenant.

The logocentrism of animal rights is expressed not only on a conceptual level, but also on a practical one. Specifically, the enforcement of animal rights illustrates the problem of logocentrism in praxis. In "Can Animals Sue?" Cass Sunstein points out that even though anticruelty laws do exist to protect animals, these laws are not enforced as often as they should be. Enforcement takes place only through public prosecution, but animal protection is generally a low priority for prosecutors (Sunstein 2004, 252–53). For example, the United States Department of Agriculture (USDA) often falls short on enforcing the Humane Methods of Livestock Slaughter Act. As watchdogs of the food industry point out, there are significant conflicts of interest within the USDA—it is populated by agribusiness lobbyists who are supposed to monitor the very industry that they have promoted (and continue to promote).[21] It is hardly surprising that enforcing animal protection has not been a priority.

Furthermore, the enforcement gap is maintained in part by the fact that animals are not considered legal persons. Even though there are animal laws protecting certain animals from gratuitous cruelty, animals possess no legal rights and have no standing in court.[22] In order to close this "enforcement gap," Sunstein argues, "private citizens should be given the right to bring suit to prevent animals from being treated in a way that violates current law . . . animals should be permitted to bring suit, *with human beings as their representatives*, to prevent violations of current law" (252, emphasis mine).[23] Sunstein's solution to the enforcement gap shows that the efficacy of animal rights is predicated on one's ability to speak: even if animals are granted legal standing and the rights to sue, they still need experts to interpret and translate their voice, as well as human representatives to speak on their behalf. The voice of the animal is twice removed even when we give it a hearing.

While the primacy of "rights" in the literature is expedient for legislating against animal cruelty and institutionalizing animal protection, it comes at a cost. As critics of liberal rights theory persuasively argue, the focus on rights can obscure the power relations on which exploitation and oppression rest.[24] Catharine MacKinnon argues that granting rights to members of oppressed groups only guarantees their equality *formally*, but it does not necessarily bring about actual equality if the structure of power remains untouched.[25] Wendy Brown makes a similar point: "while [rights] formally mark personhood, they cannot confer it; while they promise protection from humiliating exposure, they do not deliver it" (1995, 127). Although MacKinnon and Brown are concerned with human rights in their respective works, the point they make is relevant to the animal rights debate. The "enforcement gap" that Sunstein calls our attention to shows precisely the discrepancy between the granting of rights and the exercising of rights. Even if animals would be granted certain rights in our legal system, these rights are not necessarily *accessible* to them. The political voice of the animal can be compromised because of its inability to speak. As a result, the extension of rights to the animal has real force only if there is a human representative to speak for the animal, and this dependence on a human representative reinforces the existing power hierarchy. As such, the enforcement of animal rights depends on, and exacerbates, the linguistic divide that has segregated humans and animals throughout the history of philosophy.

## THE PROBLEM OF SPEAKING FOR THE ANIMAL

What does it mean to *give* someone a voice? Is it possible to give someone a voice without, at the same time, robbing them of a chance to speak? Is it true that, as Patricia Williams tells us, "the best way to give voice to those whose voice had been suppressed was to argue that they had no voice" (1991, 156)? Speaking for the other is a tricky business because the other always exceeds our knowledge, and our knowledge often exceeds our words. The difficulty of speaking for the other is exacerbated when this "other" is the animal other, with whom we do not share a language. As Thomas Nagel would want us to believe, we could never authoritatively speak for a bat because we do not know what it is like to be one. Setting aside the privileging of consciousness in his argument, Nagel is quite right to point out that simply acquiring objective facts about the other does not mean we can inhabit the worldview of the other or know what it is like to be the other (1974, 438–39). Of course, the more we learn about and attend to animals, the better we know them. We also share so many qualities in common with different species of animals that it would be disingenuous to claim complete ignorance. Yet insofar as the animal is "the other," the animal always exceeds our knowledge and words. Our representation of the other, especially the animal other, is never definite or complete.

In addition to the general challenge of speaking for a radically different other, it is also important to keep in mind that animal advocates are not the only ones speaking for the animal. As laboratory veterinarian Larry Carbone reminds us, within the debate of animal research, research advocates also speak on behalf of animals in order to justify animal experimentation (2004, 4). In other words, speaking for the animal can further, rather than mitigate, violence against the animal. Just like animal advocates, research advocates also claim knowledge and expertise with regard to the desires, preferences, and interests of animal subjects.[26] Speaking for the animal is necessary for both animal liberation *and* animal exploitation, despite their conflicting goals.

## SILENCING SILENCE

But the problem of speaking for the other is more than the failure to capture the big picture or the risk of misrepresentation, the very act of speaking for others can be a form of violence, even if the intention is to liberate rather than enslave. For one thing, valorizing the act of speaking for others reflects the primacy of speaking, thereby privileging those who are in the position to speak. For another, it is more difficult to appreciate the potency of silence when we feel compelled to "give voice" to the voiceless. As we will see below, even silence conjures its own power. The impetus to speak for the other may produce the paradoxical effect of silencing silence.

As we saw, for Derrida silence can be an exhibition of power rather than vulnerability. God's silent treatment is an expression of sovereignty by virtue of God's "right" to withhold response. We do not know when—or if—God would ever respond to our plight and prayers. Yet we cannot hold God accountable because without response, there is no responsibility. In *HumAnimal*, Kalpana Rahita Seshadri offers an even more radical account of silence through her reading of Chesnutt's story, "The Dumb Witness" (2012). In Chesnutt's story, a slave woman was literally silenced by her master's nephew when the nephew brutally cut out her tongue. Unbeknownst to the nephew, however, the slave woman was the only person entrusted with the secret of the family's will, without which the nephew could not legally inherit his uncle's property, including the slave woman. Having lost her tongue, the slave woman made use of her default muteness to withhold the secret of the will. The slave woman's silence became an active resistance that rendered the will unenforceable, guaranteeing her freedom. Through her reading of the story, Seshadri argues that the silence of the slave woman is manifested as a *capability* to withhold. Silence is a power rather than a deficiency; it is "an empty space, where the regulatory power of discourse is nullified"—just as the slave woman uses her silence to nullify her master's will (35). If we follow Seshadri's analysis, silence is even more than an exhibition of the sovereign's power through the asymmetry of responsivity (as Derrida argues), for silence can even *invalidate* discourse and reverse the law. If Derrida's and Seshadri's conception of silence as an active power is right, then we risk undermining the potency of silence when we uncritically take up the task of speaking for the other.

The impulse to substitute voice for silence also betrays a failure to think beyond the voice-silence binary. Stephen Laycock speaks of this problem as follows:

> We insert ourselves at the heart of the purported alterior subjectivity, there to speak for the mute, to give voice to the silent—to give (or rather *impose*), that is, our own voice, not to offer the animate Other a vehicle whereby it may express itself. And to substitute voice for silence is the clearest demonstration that we have not yet attuned ourselves to the silence *beyond* "silence," to the fact that "silence" is still the term of a binary conceptual contrast. (1999, 277)

As Laycock rightly puts it, the problem of speaking for animals goes beyond the risk of inaccurate projection. This is because when we insist on speaking for others, we implicitly subscribe to the voice-silence binary. We sustain and reinforce the hierarchy of power that favours beings with speech—human beings.

## SPEAKING FOR THE ANIMAL OTHER AND HUMAN EXCEPTIONALISM

That speaking for the other can itself become an instance of oppression of the other is a familiar problem in feminist and postcolonial discourses. Gayatri Chakravorty Spivak famously argues that the intellectual's representation of the subaltern obscures the voice of the subaltern, while Linda Alcoff cautions against the impetus to speak for others. For Alcoff, the practice of speaking for others is often "born of a desire for mastery," and those who speak for the other often reinscribe the hierarchy that privileges the speaker in the first place (1991, 29). Accordingly, it is important for us to interrogate our own investment in speaking for others, and how speaking for others may sustain the privileged position that we assume.

The concern that Alcoff articulates has been rehearsed by animal advocates. For example, Laycock insists that speaking for animals can become a form of "paternalism," a "human chauvinism" that inhibits genuine openness to the animal other (1999, 279). Borrowing from Said's notion of Orientalism, more recently Jason Wyckoff argues that speaking for animals can be a form of "dominionism" in which animals are reduced to mere resources and objects of *human* knowledge (2015). The very act of speaking for animals can be a confirmation of our dominion over them. As such, speaking for the animal is yet another manifestation of human exceptionalism.

It is important to remember that in Alcoff's essay the other in the "speaking for the other problem" is not the animal other. Given that feminists (and postcolonialists) have interrogated the problem of speaking for the other long before the posthumanists,[27] it is not surprising that the problem of speaking for the animal is often compared to the problem of speaking for other oppressed groups. In "Of Mice and Men" (2004), Catharine MacKinnon points out that the problem of speaking for the other is a problem common to women's issues as well as animals'. She is especially concerned with the question of what counts as a legitimate solution to animal oppression and who has a legitimate *say* regarding this solution. "Just as it has not done women many favors to have those who benefit from the inequality defining approaches to its solution, the same might be said of animals. Not that women's solution is animals' solution. Just as our solution is ours, their solution has to be theirs" (2004, 270). I will return to the distinction between women's solution and animals' solution shortly. At the moment I want to focus on the idea of "animal solution." What exactly is an "animal solution"? What does MacKinnon have in mind? What (and whose) standard do we use to measure the success of an animal solution?

## ANIMAL SOLUTION: ANIMALS SPEAKING FOR THEMSELVES

Almost immediately following her call for an "animal solution," MacKinnon asks that we consider "what it would do to the discussion if [animals] *spoke for themselves*" (2004, 270; emphasis mine). This move signals a new direction in animal advocacy. Instead of speaking for the animal uncritically and imposing a human solution, an animal solution requires us to listen to the animals, recognizing animals as active participants in their liberation. Furthermore, it requires us to problematize the linguistic divide: at stake is not whether animals can speak, but whether *we* can understand them when they speak and "what it will take to learn the answer [from the animals]" (MacKinnon 2004, 270). As Kari Weil puts it, we cannot determine decisively that animals are deficient in language simply because they do not respond to us, for it could be *our* failure to give them a reason to speak (2010, 6). Similarly, Cynthia Willett argues that the human-animal divide is "grossly exaggerated by a *human failure* to pick up on animal social cues, community formations, and possibilities of solidarity" (2014, chap. 1; emphasis mine). Paraphrasing Bentham, we may say that the question is not can they reason? Nor can they talk? But can *we* listen and understand? It is not the animal's burden to prove that it is worthy of moral consideration, but our responsibility to listen carefully.

Like MacKinnon, Wyckoff also argues for the importance of listening. In response to the question of how we might responsibly advocate for animals, he remarks, "A beginning might be a rethinking of the ability of *animals to speak for themselves*, perhaps not linguistically but through non-linguistic expressions of preference" (2014, emphasis mine). We do not need propositional statements in order to express some of our basic emotions, desires, and preferences. Our body language often expresses (sometimes betrays) what we *really* think: a slight frown or a flinch when your boss puts his hand on your shoulder indicates that this is unwelcome contact, even if there is no explicit protest. Similarly, propositional claims are unnecessary for communicating the undesirability of living in an overcrowded coop or the pain of being castrated without anaesthesia. There are many ways to discern and attend to what an animal wants without the use of propositional claims. To let animals "speak" we must also reconceptualize and broaden the meaning of "speaking" to include a plurality of ways to speak.

Both MacKinnon and Wyckoff emphasize the importance of listening by insisting that we let animals speak for themselves. Given the ideological baggage that comes with speaking for the other, they see a need to change the paradigm of animal advocacy from "speaking for" to "listening to." The assumption behind this drastic shift goes like this: when we speak for animals, we are obscuring their voices and not letting them speak, thus we fail to listen when we are too busy speaking for them. "Listening" is supposed to be a remedy that curbs our impulse to speak for the other, a solution to the violence of giving a voice to the voiceless. But what does it really mean to speak for the other, to give voice to the voiceless? Does speaking for animals necessarily mean shutting down their voice and imposing our own? It seems that speaking for the other has gotten a bad rap.

The call for listening is certainly important—however, its importance is almost tautological within the context of animal advocacy. After all, isn't listening precisely what we need to do *in order to* speak for the other? Those who have taken up the task of giving a voice to the voiceless may protest against the assumption that they fail to listen simply because they speak for the animal. In all likelihood, animal advocates would tell us that they *do* listen; in fact, it is because they *have listened* to the animal's plight that they feel compelled to speak for them, to articulate for the animal what they (the advocates) have heard. Making the effort to explore and cultivate different ways of listening to animals gives us a better sense of what animals want. We become better representatives and interpreters, more capable of giving a voice, when we listen better. Given that listening is a prerequisite of representing others, the two are not mutually exclusive. Thus it is puzzling that listening is presented as a "remedy" to curb our impulse to speak for the other when the success of the latter hinges on the former. As such, even though MacKinnon changes the question from whether *they* could speak to whether *we* could listen, the need to listen to animals is perfectly compatible with the impetus to give voice to the voiceless.

That giving a voice need not mean *imposing* a human voice is particularly evident in the following example. The urgency to legitimate the animal's voice has already prompted some concrete, practical recommendations on how to incorporate the animal's perspective when writing about animal issues. In an article published in *Journalism Studies*, Freeman, Bekoff, and Bexell argue that professional journalists ought to consider the voice of animals "as a legitimate source or perspective" when covering animal and environmental issues (2011, 596). Significantly, they motivate their argument by invoking the tenet of "giving voice to the voiceless," which is in the code of ethics for professional journalists.[28] They argue that the code should be "expanded to include . . . our fellow *animal species* whose voices often go unheard"[29] (Freeman, Bekoff, and Bexell 2011, 590). Like MacKinnon and Wyckoff, Freeman, Bekoff, and Bexell acknowledge that animals do have voices—they just often go "unheard." A responsible journalist is willing to take up the task of excavating and articulating their voices for the audience.

The invocation of "giving voice to the voiceless" is important here. For Freeman, Bekoff, and Bexell, "giving voice to the voiceless" does not carry the unsavoury connotation of imposing from without, rather it means letting animals speak for themselves and listening to their perspectives. In other words, "giving voice" need not mean speaking over the other, but listening and helping the other articulate its voice. As we will see, not only is this a far more charitable way to characterize the effort to speak for the other, it is also a more honest one as it acknowledges the necessity of human representation and mediation in animal advocacy.

## THE PRIMACY OF THE ANIMAL'S VOICE

Can we really do anything for animals without, in some ways, putting words in their mouths? At this point, it is instructive to compare MacKinnon's injunction to let animals speak for themselves with the following quote by Laycock. After chastizing philosophers for "paternalism" and "human chauvinism," Laycock writes,

> It is no excuse to complain that in attending the Other's voice we hear nothing, that we must speak for the Other because the animate Other cannot speak for itself, that the screen would be blank without our own projection. Let the screen lapse into imageless blankness. *Let all lapse into silence*. (1999, 279; emphasis mine)

For Laycock, it is important that we check our impulse to speak for the animal, and like MacKinnon, he too argues that we need to attend to the animal's voice—even if that voice is delivered in the form of silence. After all, a commitment to listening does not guarantee that we hear anything, nor does it mean that animals are willing to speak to us. For Laycock, "lapsing into silence" is preferable to projecting our ideas and desires to animals, especially when listening yields nothing.

I have two concerns regarding the appeal to "lapse into silence." The first is political. What might be the material force of "lapsing into silence," especially when it comes to transforming our relationship with animals on the ground? How do we tell the difference between silence and indifference? Or even silence and quiet hostility? How do we prevent this "lapsing into silence" from lapsing into an excuse to retreat from standing up for the oppressed? Alcoff reminds us that retreating from speaking for others "significantly undercuts the possibility of political effectivity" (1991, 17). This concern is even more germane in the context of speaking for animals. As we saw, one way to close the enforcement gap is to have a human representative sue/speak on behalf of the animal—the political effectiveness of anticruelty laws hinges on a human speaking for the animal. Accordingly, we must ask ourselves: how can we stop the material violence against animals when we get caught up in the metaphysical violence of speaking for them?

Second, the injunction to let the animal speak and the valorization of the animal's voice betray a certain anxiety about projection and imposition. To see the logic of this anxiety, Carbone's observation on the tradition of academic writing is instructive.

> Scientists (and scholars in many fields) use the impersonal passive voice in their writing as a sign of their attempts to remove their particular interests and biases from their project at hand. They pose as mere bystanders, objectively reporting nothing but the facts, dispassionately explaining what those facts mean. Nature speaks through them. (2004, 16)

Scientists imagine themselves as disinterested reporters who simply "let nature speak." But the idea that their writing is merely a report on *what nature says*, free from human biases, is an illusion. Following Carbone, we may ask, are we valorizing the animal's voice because we too want to remove ourselves—as well as our anthropocentric interests—from our writings on animals? Just as scientists imagine that "nature speaks through them," are we not in some way hoping that animals "speak through" us? But if "facts" are already imbued with the interpreter's interests, then "animal voices" are also subjected to the perspectives of the translator and interpreter.

Suppose "letting animals speak" does not collapse into a disavowal of responsibility and it still calls for our response. In order to respond we must first translate the animal's voice into our terms. Similarly, "lapsing into silence" still calls for an interpretation—a human interpretation—of such a silence. In other words, *some* representation must take place if we wish to respond to the animal. Given that most of us are not attuned to the subtle and individualized ways animals communicate, we rely on experts to translate and interpret their voices for us. (Even the experts themselves have to translate the animal's voice in terms that are intelligible to themselves and other humans.) As such, even if we are willing to listen, even if we reconceptualize "speaking" to include a variety of communicative modes, the animal voice must still be mediated by a human representative.

As with human communication, translation and interpretation are integral to our listening and understanding of the animal's voice.[30] But with cross-species communication, the risk of mistranslation or misinterpretation is predictably greater. My point is not that a good translation is impossible.[31] Rather, my point is that even when we let animals speak for themselves, we cannot easily extract the animal's voice from the human's voice. There is always the risk of projection, imposition, or anthropomorphism when it comes to translating and interpreting the animal's voice. While the appeal to let the animal speak may curb our unreflective impulse to speak for the animal, it does not give us a voice that is properly animal, free from the contamination of human biases. Unlike the injunction to let animals speak for themselves, the language of *giving* a voice to the voiceless or *speaking for* the other readily acknowledges the active role humans play in representing animals. Like nature, animals do not "speak through us." Every time we represent an animal, we are constructing their needs and desires. Perhaps the best way to minimize projection and imposition is not to imagine that there is such a thing as the "animal voice" free from human prejudices, but to recognize the inevitability of human representation in animal advocacy and then work to minimize misrepresentation.

## CAN WE GET AWAY WITH *NOT* SPEAKING FOR ANIMALS?

In "How to Do Things with Animals," Karen Raber remarks: "It is the peculiar fate of animal studies that the subjects of its liberationist impulses will not speak up" (2008, 107). And as a result, "paternalism and repressive displacement" seem inevitable in our representation of animals (2008, 107). Carbone makes a similar claim regarding animal welfare policy: "What sets animal welfare policy studies apart from most other policy studies is that animals have no direct voice. They enter policy dialogues only through those people who would speak for them" (2004, 5). Historian Erica Fudge also speaks of the impossibility of writing a history of the "inarticulate" (2002, 5). Even though there is a history of human representation of animals, there is no "history of animals" because animals do not write their own history (2002, 5). And for philosophers, the question becomes this: Is a philosophy of the animal possible?[32] If a history of the animal requires documents left by the animal (Fudge 2002, 5), then does a philosophy of the animal require thoughts thought by the animal? Can we philosophize about animals without any anthropocentric investment and projection?

Although I highlighted the connection between rights and speech earlier, the speaking for others problem surely does not belong to the rights discourse alone. As long as we write and talk about animals, we are constructing their perspectives and representing their positions.[33] In fact, even the appeal that we should let animals speak for themselves involves a representation of their desire, namely, the desire that they would rather speak for themselves than have a spokesperson. It is ironic, though not surprising, that this particular representation of the animal betrays our own bias in favour of speaking for ourselves—it is important for *us* to speak for ourselves, so it must also be the case for the animal. My point is not that MacKinnon and Laycock have *mis*represented the preference of animals, or that animals would have preferred a spokesperson. Rather, my point is that representation is inevitable even in the most radical rejection of representation.

At this juncture, we begin to see where the problem of speaking for *animals* departs from the problem of speaking for *women*. And although Alcoff's treatment of the speaking for others problem has been instructive, we also begin to see its limits when this "other" is an animal rather than a woman. Following Spivak, Alcoff makes the following recommendation: "We should strive to create wherever possible the conditions for dialogue and the practice of speaking *with* and *to* rather than speaking *for* others" (1991, 23; emphases mine). Those who are in the position to speak must seek to democratize the conversation by ensuring that the "other" has a chance to speak, too. Now it seems that we have already created "conditions for dialogue" when it comes to human-animal communication: Koko the gorilla has learned how to sign, and captive dolphins have learned to read a version of the American Sign Language. But even in these relatively successful examples, the "conversation" is far from egalitarian. In the case of Koko, she was given a chance to speak only on the condition that she speaks (signs) *our* language. So a dialogue with Koko is legitimized only by the valorization of human language. Our "dialogue" with dolphins is even more precarious. As an underwater acoustic consultant points out in the documentary *The Cove* (2009), it is ironic that we teach dolphins American Sign Language—dolphins do not have hands. Communication is reduced to a one-way process: us commanding the dolphins. Surely, we may create "conditions for dialogue" not by teaching animals human speech, but by attending to their specific body languages and social habits. But in order to make sense of these body languages and social habits, interpretation (or translation) is once again necessary. At the end, the distinction between speaking *with* animals and speaking *for* animals may simply collapse.

Considering the significance of feminist issues to animal issues, MacKinnon notes: "Not that women's solution is animals' solution. Just as our solution is ours, their solution has to be theirs" (2004, 270). Her comment is instructive here. While feminists can do feminist things with words,[34] animals do not do beastly things with words. While feminists may participate in the language game of patriarchy in order to subvert it and expose its contradictions, animals don't write pamphlets or articulate their positions. To put it more simply, while women and members of other oppressed groups can, given the right conditions, speak for themselves, animals do not relieve us of the burden of speaking for them, nor do they relieve us of the resultant guilt of "human chauvinism."[35] Speaking for animals, it seems, is inevitable, even if it is problematic and distasteful. The linguistic divide reasserts itself despite philosophers' repeated attempts to downplay it.

That animals do not lift our burden of representing them also presents a significant challenge for the posthumanist project of decentring the human.[36] In response to theorists who seek to destabilize the human-animal divide from a posthumanist perspective,[37] Raber criticizes what she deems a "problematic desire" in animal studies, "a desire for a world in which humans and animals live in happy harmony without exploitation or abuse" (2008, 100). (The promotion of a vegan diet for pets is a good example here.) According to Raber, not only do posthumanists fail to interrogate their own utopianism, they also harbour a "fantasy of the post-cruel" in their utopianism, a "desire to eradicate human abuse of animals in all its forms" (2008, 106).[38] The fantasy that we can be free of cruelty is fundamentally at odds with the project of the posthumanist because "to reject cruelty [is] to reject the animal that we are—to reinstate human exceptionalism with all its attendant problems" (2008, 106). For Raber, the very disavowal of cruelty undergirds the human-animal divide insofar as we fail to acknowledge that cruelty is simply part of nature. Killing and violence are inevitable for the survival of some animals. Accordingly, the desire to reject cruelty exposes our anthropocentric desire to transcend nature and our own animality. In other words, the problem is not simply that animal advocates are too unrealistic in their utopianism, but that their utopianism unwittingly reaffirms the human exceptionalism they try to overcome.

The problems of Raber's argument are proportional to their provocativeness. However, I will not return to her argument and my critique until chapter 5. As I will argue there, Raber's argument is predicated on a dangerous assumption about animality, an assumption that feeds not only speciesism, but also racism. For the purpose of the present chapter, I simply want to draw a parallel between the fantasy of the post-cruel and the injunction to let animals speak for themselves. In a sense, this fantasy of the post-cruel is a different expression of the fantasy of having animals speak for themselves. Just as utopian animal advocates envision a world without violence and cruelty against animals, MacKinnon and Laycock imagine a world without the tyranny of speaking for others. However, just as we cannot escape human exceptionalism even in the most radical animal egalitarianism, we cannot avoid representing animals even in our disavowal of speaking for them. In her paper, Raber asks whether it is "possible that we *need* human exceptionalism" in animal ethics (2008, 101). Following Raber, we may ask whether it is possible that we *need* to speak for animals in animal ethics, and whether it is possible that we *need* the violence of "human chauvinism" or "paternalism" in order to curb the worse violence of animal exploitation.

There is, of course, more to the question of language in animal ethics than this linguistic divide. In *The Sexual Politics of Meat*, Carol Adams analyses the ways our language "upholds meat-eating" (1990, 63). In her view, our anthropocentric language alienates us from other animals. For instance, the use of "it" as a pronoun for the animal reduces the animal to a thing-like status, while different animal metaphors "distort the reality of other animals' lives" (1990, 64). Furthermore, Adams argues that "behind every meat meal is an absence, the death of the animal whose place the meat takes" (1990, 63). Once butchered, seasoned, and cooked, the animal on our plate is no longer recognizable *as an animal*; in a similar way, once baptized "meat," animals disappear via a linguistic sleight of hand (1990, 64). For Adams, the consumption of flesh is sanctioned in part through the magic of renaming.[39]

Adams helps us see a connection between language and diet, between what we say and what we eat.[40] Indeed, Adams shows that insofar as we are linguistically constituted, it is important to examine what our linguistic capacity has afforded us to do to—and do for—the animal. If my analysis is correct, the capacity to speak also entails the responsibility to speak for our animal others (whether in the form of representation, interpretation, or translation)—not because we've decided once and for all that animals do not speak or that we refuse to listen, but simply because we cannot avoid speaking for or representing them.

In this chapter I argued that the linguistic divide is still relevant in animal ethics. Insofar as speaking for the animal other is a responsibility that we cannot disavow, we need to further examine two important issues: (1) how can we speak better for the animal? or how *not* to speak for the animal, and (2) how can we acknowledge the human-animal language difference without pitting ourselves against the animal? That is, how can we live up to our status as "the speaking animal" without severing our relationship with the animal other?

I will take on the first issue in my last chapter, where I examine a particularly injurious way of speaking for the animal. As I hope to show, until we undo the problematic link between cruelty and animality, animal advocacy risks succumbing to both racism and speciesism. I will address the second issue in chapters 3 and 4, where I present an account of language that is more conducive to animal advocacy. But before I move on to chapter 2, I would like to present some preliminary remarks on the importance of reconceptualizing language.

## RECONCEPTUALIZING LANGUAGE

My analysis in this chapter is made against the backdrop of philosophers' quick dismissal of the linguistic divide as morally relevant in the literature. Their dismissal is not unreasonable: for one thing, they rightly point out that language capacity should not determine our moral status, so the divide appears to be a moot point. For another, the linguistic divide is not an innocuous divide, but a hierarchy that places human above animal. The linguistic divide is often employed to demonstrate human exceptionalism. In fact, linguistic capacity is often associated with other capacities that we typically reserve for humans, such as the capacities to reason or philosophize. If language is associated primarily with reason or rationality, then drawing attention to the linguistic divide may reinforce the human exceptionalism that animal rights philosophers seek to overturn. As such, philosophers are well motivated to demonstrate the irrelevance of the language divide in animal ethics.

I should note that not all philosophers employ the *Can-they-suffer* strategy to undermine the moral relevance of *logos*. One particularly interesting alternative is to reconceptualize the ways ethical claims are made in order to include animals in our ethical sphere. In *Interspecies Ethics*, Cynthia Willett introduces a provocative multispecies ethics that takes seriously the communicative capacity of preverbal or even preconscious agents (2014, chap. 3). Willett argues that insofar as a preverbal, non-discursive encounter is also regulatory and normative, we can make ethical claims through affective attunement. Just as infant and adult co-regulate their behaviour via proto-conversation, different species of animals can cultivate social bond and normative expectations via biosocial signalling. The *lingua franca* of an interspecies ethics can no longer be human language insofar as we need to articulate alternative modes of communicating (and responding to) needs and expectations both within and across species.

Although I agree that there is an urgent need to explore different modes of registering ethical claims, my goal in this book is different. I am not particularly interested in exploring cross-species communication or redrawing the line of language. My way of challenging the human-animal language divide is neither to argue that animals, too, can speak, nor to stretch the conception of language far enough to include affective signalling. Although this may sound paradoxical, I shall argue that the language divide need not be divisive. We can speak of our unique responsibility as speaking animals without elevating ourselves above other animals. To begin considering these possibilities, we need to reconceptualize language so that it allows us to stand together with the animal other. We need an account of language that acknowledges our connections with the animal without obscuring our differences. Instead of linking language to reason, we should think of language as a *relational* capacity: our capacity to speak and our capacity to relate are constitutive of each other.

## NOTES

1. Bentham (1996), chap. 17, section 1, footnote to paragraph 4.
2. Garrett's article (2007) on the origin of animal rights is helpful here. Having examined Francis Hutcheson's pioneering work on animal rights and its influence on Bentham, Garrett concludes: "Instead of universal rights being expanded to include all types of rational, mature human beings, the non-cognitive ability to feel pain and pleasure became the focus, moving rights beyond reason and species to include all types of pain-avoiders and pleasure-seekers" (265). Derrida also points out how Bentham changes "the question" regarding the animal in philosophy (2008, 27–28). See also Napolitano (2010, 52–54).
3. Singer (2009, 7).
4. Sorabji wittily captures the absurdity of using linguistic capacity to justify exploitation: "The Stoic retreated to the position that at least they don't have syntax. The moral conclusion was meant to be 'They don't have syntax, so we can eat them'" (1993, 2).

5. See also Warren (1997, 61).
6. For a general account on how philosophers deal with the linguistic divide, see Steiner (2005, 18ff.).
7. See Cavalieri (2001, 20); Wise (2002, 135–58).
8. See Wise (2002, 207–30); Fellenz (2007, 46).
9. Singer (2009, 14); Dombrowski (1997); Fellenz (2007, 46–47).
10. Even if our capacity to speak is linked to our status as moral *agents*, it is not linked to our status as moral *patients*.
11. For Derrida, the privileging of *logos* goes hand in hand with the human-animal divide: "logocentrism is first of all a thesis regarding the animal, the animal deprived of the *logos*, deprived of the *can-have-the-logos*" (2008, 27).
12. Weil (2010) discusses the "animal turn" in academia. Her essay offers many helpful insights on issues of representation and the problem of speaking for animals.
13. See Wolfe's (2010) analysis on rights discourse in the current framework of the animal debate.
14. Singer (2009) considers critiques of animal rights "irrelevant to the case for Animal Liberation" because his argument for animal equality does not rest on the assumption of animals having rights (8).
15. Donovan and Adams (2007) identify five major problems with the rights discourse (introduction, esp. 4–6). See also Adams (1994). Donovan argues that the rights discourse continues to rely on "mechanistic premises of Enlightenment epistemology" (2007, 69). McKenna argues that Singer's argument for vegetarianism prioritizes reason over emotion (1994), while Kheel characterizes the primacy of rights as a "masculinist orientation" (2004, 336).
16. In fact, the parallels between human and animal rights are so strong that "their universality stands or falls together" (Donaldson and Kymlicka 2011, 45).
17. Hobbes, of course, famously points out this connection in *Leviathan*, chapter 16. Following Benveniste, Bourdieu also reminds us of the connection between "*auctoritas*" and "*auctor*" in *Language and Symbolic Power* (Bourdieu and Thompson 1991, 222).
18. In addition to "authority," the word "dictatorship" also intimates the connection between power and speech. According to the OED, to "dictate" means "to put into words which are to be written down; to utter, pronounce, or read aloud to a person." But it also means "to prescribe (a course or object of action); to lay down authoritatively; to order, or command in express terms."
19. Because animals do not have language and cannot make covenants with humans, Hobbes also claims that humans cannot do injustice to animals (Francione 2000, 123).
20. It is noteworthy that animal advocates are often accused of anthropomorphism when they speak for the animals, while the term "anthropomorphism" referred initially to the act of attributing human qualities to deities.
21. Take some key players in the Obama administration as examples. Tom Vilsack, an advocate for Monsanto and genetic engineering, has served as the Agriculture Secretary since 2009; Islam A. Siddiqui, another supporter of Monsanto, was the Chief Agricultural Negotiator between 2011 and 2013; and Michael Taylor, another friend of Monsanto, has led the FDA since 2010.
22. The Nonhuman Rights Project (NhRP) has been trying to change the common law status of chimpanzees from mere "thing" to "persons." In April 2015, two chimpanzee captives in Stony Brook University were granted writ of *habeas corpus* by a New York judge when the judge ordered a hearing for them, but within a day the judge amended the order by striking the language of the writ from the order (McKinley 2015).
23. Sunstein proposes a different solution to this "enforcement gap" in a later article. See Sunstein and Leslie (2010, 122ff.). In it he suggests that we adopt "disclosure of information as a regulatory tool" (123). Such a disclosure can be done through, for example, a "labeling system" (136). Food labels are supposed to inform consumers of the practice of a given meat supplier, and the point is that "consumers [would be] empowered to make food choices that take into account their preferences for different levels of animal welfare" (136).

24. For a helpful summary on CLS scholars' critiques on rights, see Schneider (1990) and Schwartzman (1999). In many cases, critiques of liberal rights overlap with feminist critiques of animal rights.

25. This particular critique by MacKinnon is presented by Schwartzman (1999, 34).

26. As Carbone points out, "[both] animal protectionists and scientists claimed to speak for the reality of the lives of research animals" (2004, 169). For example, animal advocates describe "unspeakable tortures" in research labs, while some research advocates claim that "not all fatal toxicity studies are painful to the animal subjects" to justify animal experimentation (2004, 169).

27. There are different kinds of "posthumanists." For the purpose of this chapter, I am using this term in a general way to refer to thinkers who challenge the centrality of the human (anthropocentrism); their anti-anthropocentric sensibility often has important implications on human-animal relationships.

28. Under the principle of "Seek Truth and Report It," professional journalists are expected to be "vigilant and courageous about holding those with power accountable. Give voice to the voiceless." *Society of Professional Journalists Code of Ethics*, revised September 6, 2014.

29. This call for expansion is rather curious. On the one hand, it endorses the recent trend of shifting the focus: it is not that our fellow animal species are voiceless, but that their voice is *unheard by us*. But on the other hand, the strategy of *expansion*—it works for the oppressed humans, let's expand it to other species—is one that feminists and posthumanists have cautioned against.

30. The need for translation and interpretation is not exclusive to the human representation of animal. As Alcoff argues, each representation is "the product of interpretation" (1991, 9). Both "speaking for others" and "speaking about others" involve "the act of representing the other's needs, goals, situation, and in fact, who they are" (1991, 9).

31. Even if prejudice and distortion are inevitable in translation, some translations are still better than others.

32. In an article on elephant communication, Phillips highlights the problem of "scriptability," which involves the question of "whether [elephant] behavior . . . can be *written* (or written about) in the first place" (2010, 34).

33. Alcoff's discussion of the distinction between "speaking for others" and "speaking about others" is helpful here. She argues that the distinction is not necessarily clear-cut. For her, both practices involve "the act of representing the other's needs, goals, situation, and in fact, *who they are*" (1991, 9). Each representation is "the product of interpretation" (1991, 9).

34. Hendricks and Oliver (1999).

35. See Raber (2008, 107).

36. For a helpful discussion on the poststructuralist project of decentring the human, see Raber (2008, 99–101).

37. Wolfe describes the posthumanist project as "challenging the ontological and ethical divide between humans and non-humans that is a linchpin of philosophical humanism" (2008, 8).

38. It is interesting that, speaking of the "ferocity of man," Lacan insists that "cruelty implies humanity" (quoted in Derrida 2008, 105). In chapter 5 I will critically examine the connection between cruelty and humanity.

39. For a detailed account on how language justifies the oppression of animals, see Dunayer (2001).

40. Examining the works of Derrida and J. M. Coetzee, McKay (2010) makes the following observation, "Language use is explicitly linked to meat eating; speaking and carnivorousness are regarded as the always-already of humanity" (68). For an interesting account of the relationship between assimilation of sounds and the assimilation of flesh in Rousseau, see Oliver (2009, esp. chaps. 2 and 4).

## *Chapter 2*

# From Animal Father to Animal Mother

## *A Freudian Account of Animal Care Ethics*

Anna Freud used to write poems for her father, Sigmund, on his birthdays. These poems, however, were written on behalf of their dogs and were presented as gifts from the dogs. According to one account, "Sigmund would always read the poem out loud, with great dramatic flourishes, then thank the dog in whose name it was signed and offer the dog the first slice of birthday cake" (Coren and Bartlett 2002, 137). Aptly dubbing these poems "birthday doggerel," another commentator wonders, "Whose voice is speaking through this rhyme? A dog separated from its master? or the one-time naughty daughter expressing her love for her father?" (Molnar 1996, 277). Indeed, whose voice is speaking here? Anna the writer of the poems? Sigmund the reciter whose physical voice delivers the poems? Or the dogs from whose perspective the poems are composed?

It is fitting that the "birthday doggerel" anecdote invokes the question of *whose voice is speaking*. The overlapping of voices points to the indeterminacy of authorship—not only in the case of the birthday poems, but also in psychoanalysis more generally. During analysis the patient does most of the talking. While the patient is expected to tell her tale, her story is rarely the end of the story or the whole of the story. The story told by the patient during analysis is often symptomatic of the repressed or displaced psychic story. Freud, as we know, retells stories on behalf of his patients by reconstructing various scenes of the Oedipal drama in their lives.[1] For him, even though the free associations of the patient during analysis provide "raw material" for the suppressed story, it is up to the analyst to structure the raw material together.[2] In an important sense, the service of the analyst is premised on the fragmentation of the patient's story.

That the patient relies on the analyst to construct and come to terms with her psychic story calls into question the patient's ownership of her story. But the indeterminacy of authorship in psychoanalysis is symptomatic of the difficulty of being one's own author more generally. As Judith Butler argues, insofar as the self is always interrupted by the other, insofar as one's account of oneself is always an address to the other, there is no account of oneself that is entirely one's own making (2005). Indeed, psychic story aside, have we ever told a story that is absolutely our own? Is it ever possible to tell our own story without supplementary anecdotes from the other? We cannot, for example, tell the stories of our birth and our death. Even childhood memories are often contaminated by the memories of our parents or siblings. Insofar as psychoanalysis displaces the self in the unconscious, it makes us all wonder whether the voice that we take as our own is not already some kind of translation, some act of ventriloquism.

Keeping the question of authorship in mind, let's return to the dogs in Freud's family for the moment. Commentators have pointed out that Freud's dogs took on the role of surrogacy (Molnar 1996; Pellegrini 2009; Garber 1996). For instance, Anna's black Alsatian, Wolf, initially served as a paternal surrogate—Freud acquired it to protect Anna during her evening walks. He later acknowledged that Wolf also served as a "replacement" for his grandson, Heinele, who died from tuberculosis (Pellegrini 2009, 232). Wolf was simultaneously a replacement of the ailing father and a lost child. It is also noteworthy that the Freud family recycled the names of their dogs, so the new dogs also served as substitutes for the old ones.[3] As one reader comments, for Freud "all object-love participates in an endless chain of substitutions" (Pellegrini 2009, 244).

In a bizarre yet poetic twist, the dogs even became stand-ins for Freud to relieve him of his digestive duties (Molnar 1996, 275). Freud suffered from jaw cancer in the last years of his life, which resulted in difficulty eating. In one of her "birthday doggerels," Anna imagines the dogs eating—and digesting—on behalf of her father. They consume the birthday cake not as a treat, but as a duty; they became Freud's "prostheses" of the mouth that "tirelessly do [his] chores of chewing" (Molnar 1996, 276).

The logic of substitution is not just at work in Freud's personal relationship with his dogs. As we will soon see, central to Freud's account of animality is the substitution between the animal and the father. But before we go into that, I should note that it is difficult to give a singular, unified role to animality in Freud's corpus. In *The Interpretation of Dreams*, Freud catalogues various meanings of animals in dreams: a lizard symbolizes the prevention of the castration threat (as a lizard can regrow its tail); fishes, snails, cats, and mice are all symbols of genitals; small animals or vermin represent an undesired sibling, while "being plagued with vermin is often a sign of pregnancy" (1900, 357). Different commentators have also assigned different roles for the animal in Freud's work: some argue that Freud uses animality "to negatively cast female sexuality"; some make connections between animality and "primitivism"; and some see animality as a representation of "the lost world of senses and sensations" or an "unconscious archive of the past" (Roberts 2008, 30; Armstrong 2008, 142–43; Parikka 2010, 92).[4] Elissa Marder argues that the animal often serves as a conceptual intervention—"without the animals, Freud cannot account for the idea of death in the psyche" (2009, 12). The animal in Freud's corpus is an impressive multitasker.

Keeping in mind the varying parts that the animal plays, I am specifically interested in its role as the father-surrogate. Freud's most extensive discussion of animals is found in his writings on infantile zoophobia. From Little Hans's horse to Little Árpád's poultry, from the Wolf-man's wolves to the Rat-man's rats, many of these animals in Freud's work are father-substitutes. In a chapter titled "Freud's Zoophilia," Kelly Oliver offers an extensive treatment of Freud and the animal. In her analysis of *Totem and Taboo*, she uncovers the animal hidden behind the primal father, and argues persuasively that the animal-father substitution is central to the Oedipal drama that drives Freud's psychoanalytic theory (2009, chap. 11). Following and building on Oliver's analysis, in this chapter I interrogate the substitutive role played by the animal. But while Oliver exposes the animal-father, I am after the animal-mother. As I will show, there is an important moment in Freud's corpus in which the mother, rather than the father, is the animal. Unlike the animal-father who functions as a figure of prohibition, this animal-mother introduces a different way to consider the human-animal relationship.

In what follows, I will first give an account of the place of the totem animal in the inauguration of the paternal law. I will then examine the abolitionist approach in the animal rights movement, and how such an approach inevitably summons the animal as a figure of prohibition. Given the narrow focus of the abolitionist approach, recent works on animal ethics have expressed the need to articulate the human-animal relationship in positive terms. I take up this call by looking at the animal-mother in Freud's psychoanalytic biography of Leonardo da Vinci. I argue that da Vinci's vulture-mother functions as a source of inspiration and creativity, rather than a figure of prohibition.

## FREUD AND ANIMAL FATHERS

It seems uncontroversial—at least according to Freud—that the feared animals are always father-substitutes. In *Totem and Taboo*, Freud makes a rather wholesale claim regarding this substitution: "It was the same in every case: where the children concerned were boys, their fear related at bottom to their father and had merely been displaced on to the animal" (1913, 127–28). Indeed, it does seem to be *the same in every case*: Little Hans displaces his fear of his father onto horses and the Wolf-man onto wolves, while Little Árpád, who is afraid of poultry, proclaims unambiguously, "'My father's the cock'" (1913, 130). But how exactly does the animal come to represent the father?

In "Inhibitions, Symptoms and Anxiety," Freud argues that the fear of animals comes down to the castration complex. Psychoanalysis reveals that the phobic little boys are *specifically* afraid of having their penises bitten off by the animal (or otherwise being devoured by the animal). The anxiety of having their genitals bitten off by the animal is, according to Freud, a "distortion [of] the idea of being castrated by their father" (1926a, 108). As we know, for Freud the fear of castration is developmentally important for a boy's life. It is by the threat of castration that the boy learns to relinquish his illicit desire to be with his mother. Now given that the castration threat is actually part of a positive Oedipal experience, the fear *alone* does not count as neurotic. For Freud, the neurosis lies in the *substitution* of the father by the animal (1926a, 103). This substitution has two advantages: first, the phobic can resolve his ambivalent feelings towards his father; that is, he no longer love-hates his father, for the hatred is transferred to the animal (1926a, 125). Second, the phobic makes the threat of castration *conditional*. While the boy can't avoid seeing his father (who presents the threat of castration initially), he can avoid seeing the horse by refusing to go on the street (in the case of Little Hans) or by refusing to read a storybook (in the case of the Wolf-man) (1926a, 125–26).

Given Freud's account of animal phobias, the animal takes the place of the father and turns into a punitive figure that threatens to castrate the little boy. Indeed, the prohibitive character of the animal is also evident in Freud's account of the primal horde in *Totem and Taboo*. The totem animal replaces the primal father and becomes the prohibitive figure that institutes the two taboos in totemism: incest and murder. In her essay "Being Human: Bestiality, Anthropophagy, and Law," Kalpana Seshadri-Crooks helpfully exposes the connection between the animal and the law by articulating their intertwined origins. She argues that the human-animal division and the law are concurrently inaugurated when the brothers commit the deed:

> The newly instituted concept of murder organizes acts of killing according to the object: those that can be killed (sanctioned slaughter) and those that cannot be killed (murder). The former group is comprised of the animal, in its difference from the human, as food or as sacrificial object. (2003, 103)

In other words, at the heart of the law against murder we find the difference between humans and animals. The law against murder (or, more precisely, the *concept* of murder) is predicated on species difference—only the killing of *humans* counts as murder. The killing of animals is, as it were, merely killing. In the time before the law there was no murder, and in the time before species difference there were neither humans nor animals. Human subjectivity is acquired in our differentiation from the animal, and this differentiation informs us of what can be killed and what cannot be killed—that is, what counts as murder and what counts as killing. Insofar as the prohibition against murder (together with the prohibition against incest) is the beginning of all prohibitions, the beginning of all laws, the totem animal is, right from the beginning, the guarantor of the paternal law.

## PROHIBITIONS AND ANIMAL RIGHTS DISCOURSE

Given the Freudian understanding of the animal as the prohibitive paternal figure, it is particularly ironic that mainstream discourse on animal rights is imbued with the language of prohibition. In his essay "The Case for Animal Rights," Tom Regan begins his argument by presenting the three main goals of the animal rights movement as follows: "the total abolition of the use of animals in science," "the total dissolution of commercial animal agriculture," and "the total elimination of commercial and sport hunting and trapping" (1985, 13).

Collectively, these goals underscore the inviolability of animal lives, and the inviolability of animal lives is motivated by Regan's deontological commitment to respect all creatures with inherent value (1985, 22ff.). Commentators have extensively critiqued Regan's position, with critiques ranging from his uncritical appeal to Enlightenment ideals to his problematic notion of inherent value.[5] I will not rehearse their critiques here. Rather, I want to draw attention to the way Regan articulates his vision of the movement. Regan expresses these goals in exclusively negative terms—he calls for the "abolition," "dissolution," and "elimination" of various practices. (In his later work Regan aligns himself with the "abolitionist" position.[6])

What does the negative prescription suggest? Why frame the goals of the animal rights movement in terms of what we are not allowed to do to animals? Surely it highlights the objectionable character of animal exploitation and the importance of halting such exploitative practices. However, Regan's negative prescription also defines the goals of the animal rights movement in terms of *prohibition* exclusively—all three goals have to do with what we need to *stop doing* to animals. The negative phrasing of his goals betrays an important limitation in Regan's vision: although it aims to radically revamp our treatment of animals, it remains reactionary insofar as it does not go beyond negating the status quo. That is, while Regan's vision offers a prescription of do-no-harm, it does not articulate how humans and animals may interact in any meaningful and positive ways. And oddly enough, by framing his vision of animal rights in the language of what-we-ought-not-do, Regan has summoned the animal as the figure of prohibition. The psychoanalytic story wherein the animal serves as the guarantor of the law finds a strange reincarnation in Regan's discourse on animal rights. That is, in his effort to plead for the animal, Regan appeals to the institution of law, an institution whose very origin invokes the figure of the animal as the prohibitive, fearful father.

It is not by chance that the prohibitive, fearful animal-father is invoked in discourse on animal rights. In *The Beast and the Sovereign*, volume 1, Jacques Derrida helpfully articulates the significance of fear in relation to the law and the sovereign; he calls this fear "the political passion par excellence" (2009, 39). The discussion of fear is brought up in the context of terrorism. Derrida exposes the "terror" that is operative in the state (or the sovereign) in order to challenge the oppositions between terrorists and the state, anti-state terror and state terror (2009, 39). He first reminds us that the name "terrorism" came from "the Terror of the French Revolution, of a Revolution that was also at the origin of all the universal declarations of human rights" (2009, 39). In other words, the historical origins of human rights—the rights after which Regan models his animal rights—are intertwined with fear and terror. Derrida then points out that in Hobbes's *Leviathan*, individuals are motivated to obey the law *out of fear*. The law is instituted out of fear (of losing one's life and property), and the law is sustained out of fear (of punishment).[7] Fear is that which sustains the power of the sovereign, insofar as the Leviathan "runs on fear and reigns by fear" (2009, 40).

Importantly, protection is promised as long as individuals submit themselves to the sovereign/the law. Derrida points out that protection has also become a form of subjugation: "'I protect you' means, for the state, I oblige you, you are my subject, *I subject you*. Being the subject of one's fear and being the subject of the law or the state . . . are at bottom the same thing" (2009, 43; emphases added). As such, protection is always an exchange, a bargain; it is offered only to the subjects who subject themselves to the law and the sovereign (and to their own fear).

That "protection" serves as a machine of subjugation is especially significant when we consider the language of "animal protection" that pervades the discourse on animal rights. What are we asking in exchange when we offer animals "protection"? In what way have we *subjected* animals by offering them our protection? Consider, for example, the "peacock bill" (SB 1533) that Hawaii passed in 2011.[8] This bill was introduced in response to a brutal peacock killing that took place in 2009. A Hawaiian resident killed an "annoying" peacock with a baseball bat and was subsequently acquitted after her lawyer convinced the jury that the peacock was a "pest." ("Pests" are normally excluded from anticruelty protection.) While an early version of the bill clarifies that peacocks are not pests, the final version of the bill merely states that the killing of the pest should be "handled in accordance with standard and acceptable pest control practices and all applicable laws and regulations." As Stephan Otto puts it, "an annoying peacock could still arguably be considered a 'pest,' however anyone considering killing this 'pest' would not be able to use a baseball bat in doing so" (2011). Although the "peacock bill" aims at extending protection to pests from unnecessary pain and suffering, it actually affirms that "annoying" peacocks can be punished by death. By "protecting" peacocks from death-by-baseball-bat, the "peacock bill" becomes a means to regulate the death of pests. As such, this bill reinscribes our sovereignty over animals by subjecting their life and death to our jurisdiction.

Let us return to fear and the rights discourse. Given the centrality of fear and terror in the sovereign power, it should not surprise us that Regan articulates his goals in terms of prohibition, and that the name "abolitionism" becomes common parlance within the animal debate. The abolitionist approach forms a great alliance with the legal discourse that pervades the literature. After all, punitive measures discourage exploitative practices that abolitionists seek to eradicate. Furthermore, the language of prohibition makes intelligible the ethical question at stake. Specifically, thinking of animal ethics in terms of what we can (and cannot) do to animals translates the rather vague notion of "human-animal relationship" into *legal* terms. Recall from Sunstein that we can now ask: How should we enforce anticruelty laws? Can animals sue? Can humans sue on behalf of the animal? The language of prohibition goes hand in hand with the emphasis on animal law in the literature. Mainstream animal advocacy often revolves around the legal status of animals—Do they have rights? Are they persons or property? Questions regarding the *moral* status of animals are often used in service of the *legal* status of animals.

Nonetheless, is prohibition a sufficient or even a necessary condition to motivate a promising human-animal relation? Should we have an animal ethics based solely—or even primarily—on prohibition? Cora Diamond is one of the earliest critics of the rights approach in animal ethics. Recall she argues that the reason we don't eat other people is not that they have a right not to be eaten, but that we don't consider them food. In her view, it is only when animals are recognized as our "fellow creatures" that they would no longer be "edible" or "food" to us. This change in language is significant. As Deane Curtin points out, the rights approach in animal and environmental ethics is "adversarial" in nature (1991, 64); it highlights what we owe to animals in the form of *non-interference*, with the assumption that prohibition is the primary way of regulating the human-animal relationship. "Fellow creatures," however, recognizes the possibility of developing friendship with animals, with the acknowledgment that our sense of fellowship also informs and governs our behaviours.

More recent works in animal ethics also explicitly challenge the focus on prohibition. We saw that Sue Donaldson and Will Kymlicka fault animal rights theory (ART) for putting too much emphasis on the negative rights of animals. They argue that this emphasis has "[deprived] ART of a positive conception of human-animal interaction" (2011, 9). In *Interspecies Ethics*, Cynthia Willett challenges the typical utilitarian can-they-suffer question that has dominated Anglophone animal ethics, as well as the Derridian-inspired response ethics that prioritizes vulnerability in the Continental tradition. She suggests: "rather than listening for the sad lament of nature, we turn to the sometimes playful, sometimes subversive social exchange within and between species that suggest a missing moral link and support a politics of solidarity" (2014, chap. 1). When we focus too much on animals' suffering and vulnerability, we risk flattening the animal to a mere moral patient to be protected. As such, there is an urgent need to articulate the human-animal relationship in positive terms.

## FREUD AND DA VINCI'S ANIMAL

Given what we have seen so far, Freud seems to be an unlikely resource in considering an animal ethics beyond prohibition. After all, animals in Freud's work are often linked to the punitive father. In Freud's account of the primal horde, the animal-father substitution plays a crucial role in anchoring the familial, social, and religious structure of a patriarchal society. The totem animal has to be the father and not the mother. In fact, the rigid substitution between the father and the animal sometimes comes at the expense of the mother. Throughout Freud's account of the totemic origin, the significance of the maternal figure is repeatedly dismissed. For example, matriarchal society is mentioned merely in passing, as Freud speculates that both homosexuality and matriarchal institution might be originated during the in-between time after the brothers' expulsion from the horde and before the murder of the father (1913, 144). When it comes to mother-goddesses, Freud acknowledges that his account of religion fails to take them into consideration, but he quickly moves on to discuss the role of the father-gods, leaving this wrinkle of his theory untouched (1913, 149).

But this animal-father substitution is perhaps not as certain as Freud would have it. In her essay on Little Hans, Anna Ornstein argues that it is not entirely obvious that Little Hans was afraid of his father rather than his mother, even though "Freud obviously maintained a deep conviction that a horse could only represent a male and, if the child was afraid, he could only be afraid of his father, not his mother" (1993, 95). As Ornstein points out, Freud insists on the animal-father substitution even when:

> Hans called the coal-carts and furniture vans the horses were pulling the "stork-box carts," which were obvious references to his pregnant mother . . . or when Hans related his fear of the horse making a row and screaming to his sister's screaming "when Mummy whacks her on her bottom and she makes such a row with her screaming." (Ornstein 1993, 95–96)

Furthermore, Little Hans drew a parallel between beating a horse and beating his mother. After Little Hans told the story of him (Little Hans) beating a horse in Gmunden,[9] his father asked him, "Which would you really like to beat? Mummy, Hanna, or me?" (What a *loaded* question.) And Little Hans immediately declared that he would like to beat his mother—with a "carpet beater" (Freud 1909a, 81).[10] More importantly, it was the mother who made the castration threat: when Little Hans was caught playing with his penis, Little Hans's mother said to him, "If you do that, I shall send for Dr. A. to cut off your widdler. And then what'll you widdle with?" (1909a, 8). (To which Little Hans replied, "With my bottom" [1909a, 8].) Given these remarks, it seems that Little Hans's mother also served as a prohibitive figure, which goes against Freud's claim that the feared animal must be a father-substitute *in every case*.

It is also noteworthy that Freud's first case of animal phobia actually concerns a forty-year-old *woman*, Frau Emmy von N. (1893).[11] While Frau Emmy's case offers many points of comparison, Freud systematically fails to make reference to her case in his examinations of infantile animal phobias (all of them come *after* the Frau Emmy case). First, Frau Emmy had an animal hallucination that seemed to be inspired by a storybook. Freud should have made a connection in his diagnosis of the Wolf-man dream, given that he insists that the Wolf-man's dream came from a fairy tale. Second, Frau Emmy's hallucinations often involved mice or rats, but Freud makes no mention of her hallucinations in the Rat-man case. Third, one of Frau Emmy's symptoms involved making a "clacking" sound that Freud refers to as a "*tic*," a sound that "resembled the cry of a capercaillie" (1893, 31). This comment is strikingly similar to the way Freud describes Little Árpád's speech inhibition, that he "abandoned human speech in favor of cackling and crowing" (1913, 130). In both cases, the animal phobics had their human speech interrupted by the voice of an animal, but Freud seems completely oblivious to this similarity when he discusses Little Árpád's case. Finally, at one point Frau Emmy told Freud, "A whole lot of mice were sitting in the trees"—a hallucination that parallels the Wolf-man's dream of the wolves sitting on a big walnut tree. Again, Freud draws no comparison to Frau Emmy's case in his analysis. Surely at the time of her treatment Frau Emmy was no longer a child, so her animal phobia should not have counted as an instance of infantile zoophobia. But how did her story get obliterated to the point of never being mentioned or referenced in Freud's analysis of animal phobias? Perhaps Freud neglected his case study of Frau Emmy for the same reason he disregarded some of the crucial claims made by Little Hans: the animal must take the place of the father and stand as a figure that issues the castration threat.

It is against the backdrop of this seemingly rigid link between the animal and the father that I turn to Freud's *Leonardo da Vinci and A Memory of His Childhood* (1910). Many pages of this psychoanalytic biography are devoted to a dream that da Vinci had as a child, a dream in which a vulture visited him. Freud tells us that the vulture is a hermaphroditic creature: it has a female body while possessing a phallus. The ambiguity of the vulture's sex is suggestive, especially in light of the link between the animal and the father that we see in cases of infantile zoophobias and the story of the primal horde. But what does the vulture substitute in da Vinci's case? What is the relationship between da Vinci and his animal? This ambiguous hermaphrodite opens up a place for us to reconsider, to challenge even, the link between the animal and the father.

## DA VINCI AND THE "KISSING VULTURE"

Da Vinci's memory of the vulture's visit, which Freud quotes from a German translation[12] of Scognamiglio's transcription of *Codex Atlanticus*, which was then translated by Alan Tyson, is as follows:

> It seems that I was always destined to be so deeply concerned with vultures; for I recall as one of my very earliest memories that while I was in my cradle a vulture came down to me, and opened my mouth with its tail, and struck me many times with its tail against my lips. (1910, 82)

James Strachey, the editor of the *Standard Edition*, points out in the preface that there is a significant inaccuracy in the German translation that Freud references. Namely, the Italian word "*nibbio*" should be translated as "kite," not "vulture" (*avvoltoio*) (1910, 60–61). While Strachey attributes this point to Irma Richter,[13] this mistake in translation was discovered as early as 1923 by art historian Eric Maclagan, and Freud apparently never publicly acknowledged the mistake (Andersen 2001, 10–11). As we will soon see, Freud specifically invokes the *vulture*-goddess in Egyptian mythology in order to ground his interpretation of da Vinci, so the mistake in translation seems costly. Since Maclagan's discovery, critics of Freud have dutifully noted this mistranslation in their commentaries on the da Vinci case: some merely mention it in passing, some see it as detrimental to his analysis of da Vinci, and some go so far as to suggest that this mistake is indicative of Freud's shoddy scholarship.[14] Given that I am primarily interested in the way *Freud* (rather than da Vinci) uses the animal in his work, the specific species of the bird is inconsequential. However, this translation debate reminds us that Freud's analysis of da Vinci is subjected to the precariousness of translation, that is, the precariousness of substitution. It serves as a reminder of the significance of—and the risks associated with—the economy of translation/substitution in discourses on animal ethics.

As noted, the ambiguity of the vulture's sex makes da Vinci's case exceptional, insofar as it departs from the norm of substitution that Freud prescribes for a patriarchal culture. Despite its hermaphroditic features, Freud insists that the vulture is a *mother*-substitute for da Vinci (1910, 88). First of all, Freud points out the phonetic similarity between the name of an ancient Egyptian goddess *Mut* (who has a vulture head) and the word "*Mutter*," the German word for mother (1910, 88). But the vulture has been associated with motherhood primarily because the vulture was believed to be an exclusively female species, a species that propagates by being impregnated by the wind (1910, 89). For Freud, da Vinci's identification as the "vulture-child" is indicative of his father's absence—he was, like the vulture, a child without a father (1910, 90).

According to Freud, while *Mut* was characterized as a mother-goddess, the Egyptians usually represented it as an androgynous figure (1910, 94). The maternal figure was equipped with the symbol of paternal potency—an erected phallus (1910, 94). The androgynous feature of the Egyptian goddess also supports Freud's claim that the tail of the vulture in da Vinci's dream signifies a phallus (1910, 85, 97). The vulture is at once a mother-substitute and a phallic power. But this androgynous goddess is as puzzling as it is fascinating: how is it possible that a mother-goddess is equipped with an erect phallus, "the mark of male potency which is the opposite of everything maternal" (1910, 94)?

To resolve this apparent contradiction, Freud turns to the children. He argues that before a male child comes to a full understanding of the threat of castration, he undergoes a phase in which the phallus is compatible with the maternal figure. In fact, the child should have no reason to assume that the mother's body is different from his own; so his mother, like him, should have a penis. It is the discovery that the mother does not have a penis that makes the castration threat real and forceful. Importantly, this discovery invokes "a feeling of disgust . . . [which] can become the cause of physical impotence, misogyny and permanent homosexuality" (1910, 96). Freud calls the time before this discovery "a time when [the male child] still holds women at full value," that is, a time before the hierarchy of gender is inaugurated and accepted (1910, 96). Interestingly, gender egalitarianism is not the only kind of egalitarianism that a child seems to take for granted.[15] In different writings

Freud has also commented on a sort of human-animal egalitarianism that he finds in children. In *Totem and Taboo*, for example, Freud points out that the human-animal hierarchy is not assumed by children:

> Children show no trace of the arrogance which urges adult civilized men to draw a hard-and-fast line between their own nature and that of all other animals. Children have no scruples over allowing animals to rank as their full equals. Uninhibited as they are in the avowal of their bodily needs, they no doubt feel themselves more akin to animals than to their elders, who may well be a puzzle to them. (1913, 126–27)

Elsewhere, Freud makes a similar observation:

> Such a displacement [in this case the displacement of the father by a horse] is made possible or facilitated at "Little Hans's" early age because the inborn traces of totemic thought can still be easily revived. Children do not as yet recognize or, at any rate, lay such exaggerated stress upon the gulf that separates human beings from the animal world. (1926a, 103)

Is the parallel between gender egalitarianism and species egalitarianism in children a mere coincidence? Do children "discover" that women are not of the same "full value" as men at the same time they distinguish themselves from animals?[16] Interestingly, the kind of egalitarianism that a child purportedly takes for granted seems to be motivated by his narcissism. A child holds his mother "at full value" based on the assumption that *like him*, his mother also has a penis. And in the case of species egalitarianism, the child ranks animals as his equal by virtue of their *likeness* in the "avowal of their bodily needs."

This "egalitarianism" based on sameness is often replicated in contemporary literature on animal ethics. In their effort to extend membership of the moral community to animals, philosophers often identify morally relevant qualities that are *shared* by both humans and animals to make their case, such as the capacity to feel pain (Singer) or being an experiencing subject of a life (Regan). But grounding animal rights on the notion of sameness can be problematic. As we saw in chapter 1, Catharine MacKinnon reminds us that "seeking animal rights on a 'like-us' model of sameness may be misconceived" because we "[miss] animals on their own terms" (2004, 264). Indeed, we may even say that an animal ethics grounded in the model of sameness is also narcissistic; and once again we see a strange parallel between Freud's account of animality and mainstream animal rights rhetoric.

In light of the connection between gender and species egalitarianism, it is telling that Freud also relegates empathy for animals to the feminine. For example, he insists that da Vinci's compassionate acts for animals—declining meat, buying birds from the market in order to set them free—are motivated by the "feminine delicacy of feeling" (1910, 69).[17] Specifically, Freud argues that da Vinci's mother's "tender seductions" were responsible for the intensity of his infantile sexual research; and the violent, sadistic traits associated with infantile sexual research were manifested in an opposite direction as an "exaggerated sympathy for animals" (1910, 132). For Freud, da Vinci's empathy for animals is symptomatic of a lack of paternal authority. Interestingly, while Freud insists that animals are father-substitutes, his interpretation of empathy for animals reflects the age-old association of the animal and the feminine. Specifically, empathy for animals is often viewed as some emotional excess that women—because they are less rational—are prone to suffer.[18]

In Freud's case studies, the relationship between the child and the animal is often ambivalent (hence replicating the love-hate feelings a child shows his father). Nonetheless, because the animal stands for paternal authority, the ambivalent feeling invariably involves fear: the Rat-man is traumatized by the "rat-punishment" (a punishment which involved putting rats into one's anus);[19] the Wolf-man is afraid of being devoured by wolves; and for Little Árpád the fowl represents a constant threat of castration. In da Vinci's case, however, the vulture does not present any obvious threat, and there is no evidence that da Vinci was vulture-phobic in Freud's account. Indeed, throughout his analysis Freud describes da Vinci's obsession with the animal as "vulture phantasy" instead of "vulture-phobia." But if da Vinci's story is not a case of animal phobia, what does it tell us? How might it transform the way we conceive of animality? Let us turn to da Vinci's dream once more.

Recall in his dream da Vinci was struck by the vulture repeatedly. The striking of the vulture may seem violent at first glance. But Freud offers two interpretations for this scene, both of which make it difficult for us to read physical violence into it. He first interprets the striking of the mouth as "an act of *fellatio*" (1910, 86), but later translates this same scene as a mother pressing "innumerable passionate kisses on [da Vinci's] mouth" (1910, 107). While both readings make explicit the sexual undertones of the dream, neither presents the vulture as menacing (if anything, the vulture is either passionate or affectionate). Also, unlike most other case studies of infantile zoophobia, the threat of castration is not mentioned. Both interpretations seem to suggest that the vulture is an object of fantasy instead of an object of fear.[20]

For Freud the tenderness of the mother is damaging. "The violence of caresses," as Freud calls it, is insidious insofar as it has "robbed [da Vinci] of a part of his masculinity" (1910, 115–17). (Da Vinci's supposed sexual abstinence is still a form of castration.) But what exactly is this masculinity of which he was robbed? As far as Freud sees it, it all comes down to da Vinci's alleged inability to pursue heterosexual relations. What the two interpretations share in common is the implication that da Vinci was a homosexual.

When Freud first interprets the dream as a fantasy for fellatio, he finds it "strange" that the fantasy is "so completely passive in character" and that "it resembles certain dreams and phantasies found in women or passive homosexuals" (1910, 86). Moreover, in the biography, Freud explicitly draws a connection between homosexuality and maternal influence—one becomes a homosexual through one's erotic relation with the mother (1910, 98–100). In *Totem and Taboo*, Freud even speculates (albeit merely in passing) that homosexuality and matriarchal society share a common origin (1913, 144). Hence, according to Freud, even though the vulture did not threaten to castrate da Vinci, it mutilated his manhood.

Freud's view of violence, of course, presupposes the norm of heterosexuality prescribed by patriarchy, and a rereading of the dream will call for resistance to this particular notion of violence. Indeed, if we rethink the hierarchy of masculinity and femininity, the kisses of the vulture-mother can be enabling rather than crippling.

It is significant that Freud identifies maternal influence as the cause of homosexuality. Recall that Freud also holds maternal influence responsible for da Vinci's "exaggerated sympathy for animals" (1910, 132). It is not a coincidence that the mother is the culprit for both homosexuality and empathy for animals. For Freud, maternal influence supposedly erodes masculinity, and homosexuality and vegetarianism are often viewed as signs of defective masculinity. Outside of the psychoanalytic tradition, feminist thinkers such as Carol Adams and Marti Kheel have called attention to the relationship between meat eating and masculinity.[21] They argue that the consumption of animals is bound up with the consumption of women; hence when a man refuses to consume meat, it is suspected that he may refuse to "consume" women as well. In her study on gender disparity in the animal rights movement, sociologist Emily Gaarder points out that "some male activists experienced ridicule relating to their masculinity and sexuality. One man described how hunters called him an 'animal rights queer' during a protest against hunting" (2011, 107).[22]

Patriarchy (whose origin is grounded in the logic of the animal-father substitution) has prescribed a specific norm governing our relationship with animals, a norm that replicates the father-son relationship in the myth of the "band of brothers." Just as the brothers in the myth see their father as a rival, we define ourselves *in opposition to* the animal; just as the band of brothers commemorate their "triumph over the father" by instituting the totemic meal, we affirm our superiority with a full-scale exploitation of animals, an exploitation that is repetitive and expansive. And just as the "savages" forgot the origin and significance of the totemic meal, our exploitation of animals has also become a mindless and institutionalized repetition. As such, the norm governing our relationship with animals in a patriarchal society replicates the father-son dynamic that grounded the paternal law. Accordingly, a deviant model of the human-animal relation (one that is based on compassion instead of fear, for example) is often relegated to the maternal in order to preserve the integrity of the patriarchal norm.

In what ways do animals pose a threat to us? Of course, animals can physically harm us. The bars through which we look at zoo animals serve as a constant reminder of the danger they pose. And even if they are not the big bad beasts that threaten us with their maws or talons, they can still threaten our health by spreading dangerous contagions—the bird and the swine seem to be the experts at that. But beyond the physical threat, animals threaten our human identity. Many qualities that we considered uniquely human—rationality, morality, language—are now purportedly discovered in various animal communities. It is no longer easy to locate human uniqueness via comparison with animals. In response to the threat of animals, we domesticate them in zoos, farms, and canine training schools and we dismember them via castration (to "fix" them, we say), experimentation, and butchering. While we may not—and probably should not—erase our fear of animals, is it possible to mitigate it? Can we imagine a human-animal relation that is not grounded in fear but in fantasy?

## A FREUDIAN CARE ETHICS

In the opening chapter of *In a Different Voice* (1982), Carol Gilligan argues that the propensity in psychology to take male experience as the norm can be traced all the way back to Freud. According to her, Freud, "who built his theory of psychosexual development around the experiences of the male child that culminate in the Oedipus complex," epitomizes psychology's male bias (1982, 6). Given that Freud was the sacrificial father in the inauguration of feminist care ethics, the two seem to make pretty strange bedfellows. Nonetheless, if we (and even Freud himself) take da Vinci's story seriously, there is a Freudian-inspired animal care ethic to be had. So while there is an extensive literature on animal care ethics, this chapter takes a psychoanalytic perspective.

As we know, ecofeminists are eager to depart from the traditional focus on rights in animal ethics. Rather than furthering an adversarial human-animal relationship, they seek an ethic that is grounded in affective qualities such as care, love, empathy, and compassion. Of course, the language of "care" can still be problematic. As disability theorist Sunaura Taylor cautions, "being 'cared for' can be stifling, if not infantilizing and oppressive" (2014, chap. 6). Furthermore, as long as the "caring for" operates in a single direction (i.e., human caretakers having custody of the animal), the protector-victim dyad presupposed by the rights discourse persists. From a disability studies perspective, Taylor argues that "being interdependent is about recognizing that we are all vulnerable beings, who during our lives go in and out of dependence, who will be giving and receiving care" (2014, chap. 6). She reminds us that humans and animals are *co-dependent*; animals "contribute in countless calculable and incalculable ways, from fertilizing the soil to offering care and friendship" (2014, chap. 6). One can still be a caretaker even when one is receiving care from others; similarly, animals can care for us even when they are dependent on our care. We need an animal care ethic that not only acknowledges, but also celebrates, the possibility of animal caring for us.

So what does Freud have to do with such a care ethic? In my view, da Vinci's relationship with the vulture-mother captures an important way in which the animal cares for us. The vulture-mother was the source of inspiration for da Vinci—it cared for da Vinci by nourishing him intellectually and artistically. Could it be that the animal cares for us by unleashing our creativity? After all, animals are everywhere on our intellectual landscape: we can find animals in novels, poems, films, children's stories, and even the Bible. It is perhaps not a coincidence that the primary subjects of our earliest cave paintings are animals, in particular big mammals. The oldest figurative cave painting is a female pig-deer, found on the island of Sulawesi in Indonesia. Art (along with technology) is often considered a uniquely human endeavour. But if art is that which distinguishes culture from nature, humans from animals, then these cave paintings suggest that perhaps it is the animal that inspires us to become human.

In addition to caring for da Vinci, the vulture-mother exemplifies another important commitment in ecofeminism—a resistance to binary thinking. Even though Freud insists that da Vinci's vulture is a mother-substitute, it is worth remembering that in Freud's initial description the vulture is a *hermaphroditic* creature: it has a female body and a phallus. Freud insists that the vulture is a female possessing a phallus, rather than a male possessing a female body. But why isn't the vulture *both* male and female? Why isn't it both a father and a mother to da Vinci? In Freud's original description, the vulture's sex is ambiguous and undecided. This ambiguity brings us back to care ethics. Given that care ethics is often presented as an *alternative* moral framework to the traditional justice-oriented moral frameworks, one common charge against care ethics is that it valorizes care at the expense of justice. In light of this charge, recent works in ecofeminism have developed and refined the relationship between care and justice. While care ethics is grounded in affective qualities such as compassion and empathy, these qualities need not be antithetical to reason, justice, or reflective judgment. For example, Deane Curtin puts forth "an ethic of compassion" in which "feelings and reasons blend and result in a moral practice" (2014, chap. 2). Similarly, Lori Gruen develops the concept of "entangled empathy" that incorporates both empathy and cognition (2015). We are capable of empathizing with others because we are always already in relation (entangled) with others. But entangled empathy is more than just a feeling, rather it is a *process*:

> The wellbeing of another grabs the empathizer's attention; then the empathizer reflectively imagines himself in the position of the other; and then he makes a judgment about how the conditions that the other finds herself in contribute to her state of mind or wellbeing. The empathizer will then carefully assess the situation and figure out what information is pertinent to empathize effectively with the being in question. (2015, chap. 2)

In order to empathize effectively, we need to reflect, judge, and eventually act. As such, while entangled empathy begins as a feeling, it has a cognitive component (reflection), a rational component (judgement), and an ethical upshot (action). For both Curtin and Gruen, affect and reason are not mutually exclusive: affect should be reflectively cultivated, and rational judgement should be grounded in feeling.

The exceptionality of da Vinci's animal—its identification with the maternal—interrupts the animal-father substitution that we see in other cases of animal phobia. Da Vinci's relationship with the vulture-mother invites us to renegotiate our relationship with animals. Specifically, it invites us to consider a maternal animal ethics that emphasizes fantasy and love rather than fear and dominance. That a human-animal relationship *can be* based on love and fantasy is certainly not something unusual. Many people identify themselves as "animal lovers," and pets are often considered family members. The question here is why such a model of human-animal relationship remains marginalized while the exploitative model of human-animal relationship remains operative. In fact, even commonly sanctioned pet-love is at times pathologized or censured: the stereotype of the "crazy cat lady" has become such a popular icon it gets its own action figure, while the image of Paris Hilton and her "purse dog" feeds the idea that pet-love is a mere accessory. If even pet-love is marginalized as either pathological or vain, we are still a long way away from a normative human-animal relationship that is grounded in love.

Although Freud pathologizes da Vinci's vulture fantasy, such pathologization is very much informed by his patriarchal thinking. Accordingly, a revision of human-animal relations goes hand in hand with the remaking of gender relations (perhaps even a remaking of the myth of the band of brothers). Freud's account of the vulture-mother remains instructive insofar as it showcases a human-animal relationship that is based on love and fantasy, instead of fear and dominance. Just as the vulture-mother kisses the infant da Vinci in his dream, intimacy can also ground human-animal relations. And just as da Vinci sublimates his desire for his vulture-mother into artistic creativity, we are also inspired by animals in our various scientific and artistic endeavours. Recognition of the animal as our nurturing mother, it seems, would point to a more nurturing (and less violent) human-animal relationship.

## NOTES

1. In *A Thousand Plateaus* (1987), Deleuze and Guattari critique Freud's interpretation of the Wolf-man's dream. They fault Freud for radically changing the content of the Wolf-man's dream in his interpretation.

2. For example, Freud uses the phrase "raw material" to refer to the Wolf-man's dream (1918, 34). In *The Question of Lay Analysis*, Freud uses the analysis of an ore to describe the materials provided by the patient during the analysis (1926b, 219).

3. According to Coren and Bartlett, "there are a number of Luns (named after their first dog), a few Tattouns (named after a favorite of Marie Bonaparte), and several generations of dogs named Jofi" (2002, 138).

4. Carrie Rohman also argues that animality for Freud is the "archaic past": it is a regressive stage that one must overcome in order to "become and remain human" (2009, 23).

5. See, for example, Donovan (1990); Oliver (2009, chap. 1); Lee (2008); Taylor (1987).

6. See, for example, Regan (2003, 1; 2004, *xiii*). In *Animal Rights, Human Wrongs*, Regan rewords his vision as follows: "The total abolition of commercial animal agriculture," "The total abolition of the fur industry," and "The total abolition of the use of animals in science" (2003, 1). Still expressed in exclusively negative terms, the rewording—now with the repetition of the phrase "total abolition"—clearly aligns Regan with abolitionism. The abolitionist approach is in contradistinction to the welfarist or protectionist approach.

7. Derrida also discusses the way fear motivates crime, so "fear is thus both the origin of the law and of the transgression of the law" (2009, 41). While the double function of fear is interesting, I am primarily interested in the way fear and terror operate in sustaining the law.

8. This bill has been effective since July 1, 2011. See http://www.capitol.hawaii.gov/session2011/lists/measure_indiv.aspx?billtype=SB&billnumber=1533.

9. A place where Little Hans's family stayed for the summer holidays.

10. In a parenthetical note, Little Hans's father added that the mother was the one who often threatened to beat Little Hans with a carpet beater (1909a, 81).

11. See "Case Histories: 2."

12. Wayne Andersen argues that Freud did not actually copy from the German translation by Marie Herzfeld, as Strachey suggests, but instead drew from more than one source "without having read Leonardo's words" (2001, 23).

13. See "Editor's Note" (61n1).

14. Andersen carefully chronicles the ongoing debate in the literature in his 2001 book (10ff.). In his 2005 essay, De La Durantaye argues that Nabokov's dismissive opinion of Freud (and psychoanalysis more generally) is grounded in Freud's lack of attention to details: "What Nabokov very consciously sought to counteract were approaches to art that, in their aspiration to uncover the general, neglected the particular. And this he found in Freud" (68). And the prime example De La Durantaye uses to buttress this observation is precisely Freud's use of the mistranslation of "*nibbio*" as "vulture."

15. That is, if we consider the absence of hierarchical thinking a negative form of egalitarianism.

16. Outside the psychoanalytic tradition, feminist thinkers such as Carol Adams and Karen Warren have pointed out the intertwined relationship between the denigration of women and that of animals/nature.

17. Andersen disputes the claims that da Vinci was a vegetarian or animal lover (see 2001, 139ff.).

18. As Emily Gaarder points out, "The image of animal rights still suffers from stereotypical portrayals of overly emotional and irrational activists" (2011, 11).

19. "Notes Upon a Case of Obsessional Neurosis" in *Two Case Histories* (1909b, 166).

20. It is not that phobia and fantasy are necessarily incompatible, but in Freud's account, da Vinci exhibits no fear of the vulture.

21. Adams (1990); Kheel (2004).

22. In light of this, it is particularly interesting that in 2010, an animal rights group was denied a spot in Sydney's Mardi Gras parade because it was "not queer enough." See Holman (2010).

# *Chapter 3*

# The Poverty of Kinship

## *Heidegger on the Human-Animal Linguistic Divide*

In "A Dialogue on Language between a Japanese and an Inquirer," Heidegger questions whether it is possible (and necessary) to consider the nature of Japanese art by way of European aesthetics. The translatability between Japanese and European arts prompts the question of language, specifically the question of translation:

> *I*: Some time ago I called language, clumsily enough, the house of Being. If man by virtue of his language dwells within the claim and call of Being, then we Europeans presumably dwell in an entirely different house than Eastasian man.
>
> *J*: Assuming that the languages of the two are not merely different but are other in nature, and radically so.
>
> *I*: And so, a dialogue from house to house remains nearly impossible. (1982, 5)

While this chapter is not about aesthetics or Japanese language, it is invested in the question regarding the possibility of having "a dialogue from house to house." In "Letter on Humanism," Heidegger speaks of language as a place in which man "dwells"—indeed, as the "*home* of man's essence" (1993, 237; emphasis mine). The imagery of a "house" is intriguing: what makes this "house" a *home* rather than a mere shelter? Who are the ones we share this house with? And who are the ones we keep out? Furthermore, if a dialogue between a European house and an East Asian house "remains nearly impossible," what can we say about a dialogue between a human house and a nest, a lair, or an animal shelter? Keeping these questions in mind, Heidegger's philosophy of language can inform the question of the human-animal divide.

At the end of chapter 1 I proposed that language appears to be a *divisive* capacity because it is often linked to reason. It is not uncommon for us to think of language as a capacity that allows us to reason, to discourse, to philosophize—pursuits that we typically regard as exclusively human. Insofar as language capacity is tied to these pursuits, it deepens the human-animal division. This explains the motivation to dismiss the relevance of linguistic capacity in animal rights discourse: it is not just that one's capacity to speak is morally irrelevant, it is also that linguistic capacity is associated with other capacities that reinforce human exceptionalism. I suggested that we need to reconceptualize language capacity: instead of linking language to reason, we should uncover the relational possibilities at the core of language.

At first glance, Heidegger may—like Freud—seem an unlikely ally for this project: his conception of animality continues to endorse the human exceptionalism that has pervaded Western philosophy. However, Heidegger is also one of the few philosophers whose account of language does not bow to the authority of reason. In fact, he is critical of the primacy of reason in our understanding of language: in "The Restriction of Being," he seeks to recover the bond between *phusis* and *logos* by "[freeing] ourselves from the opinion that *logos* and *legein* originally and authentically mean thinking, understanding, and reason" (2000, 130)[1]; in "Letter on Humanism," he maintains that the Latin rendering of "*zoon logon echon*" as "*animal rationale*" is not an innocuous translation, but a "metaphysical interpretation" of the Roman humanist (1993, 226).[2] Using Heidegger as my touchstone, in this chapter I seek an alternative account of language that gives kinship its proper place.

## HUMAN-ANIMAL LINGUISTIC DIVIDE

In his 1929/1930 lecture course *The Fundamental Concepts of Metaphysics* (*FCM* hereafter), Heidegger examines animality extensively.[3] He aims to establish three theses: (1) the stone is worldless; (2) the animal is poor in the world; and (3) man is world-forming (1995, 177). Commentators have offered extensive and critical treatments of *FCM*: some focus on the significance of temporality, historicity, and finitude;[4] some situate Heidegger's humanism within the larger context of the meaning of being;[5] while some centre on Heidegger's rejection of humanism and how this rejection remains anthropocentric.[6] Indeed, the various ways Heidegger articulates the difference between animals and humans all seem to highlight the uniqueness of human *Dasein*: animals are "captivated" by their instinctual needs while humans are opened to beings *as such*; animals merely "behave" while humans "comport" or "act"; animals are poor in the world while humans are world-forming (cf. 1995, chap. 4, especially §58 and 59). In the final chapter of the lecture course, Heidegger reveals that the animal's captivation (and thusly their capacity to "behave" and their poverty in the world) ultimately results from their lack of language. Drawing on Aristotle's notion of *logos*, Heidegger anchors his argument of world-hood to the language divide that has long segregated humans from animals.

Given Heidegger's project, it seems that a reasonable way to approach his account of animality is to articulate his conception of *logos*.[7] The question of *logos* is bound up with Heidegger's interrogation of being, thus it occupies an important place in his thoughts.[8] However, precisely because of his extensive treatment of the *logos*, it is difficult to present his view adequately within the scope of this chapter. Furthermore, it is problematic to speak of a "Heideggerian conception of *logos*" in large part because he explicitly cautions against the temptation to "reduce the nature of language to a concept"—that is, to essentialize language (1971, 188). So given that the purpose of this chapter is to develop a relational account of language, my strategy here is to examine various important moments in Heidegger's interrogation of *logos*.

In *Being and Time*, Heidegger identifies communication as one of the structural components of "assertion" (statement) (1962, §33). Although this suggests an important link between communication and language, the relational character of language cannot be established by appealing to its communicative function alone. That is, we cannot simply stretch the notion of language far enough to include all communicative gestures. This is so because, for Heidegger, there are two forms of communication: one is proper to language and the other is merely a degenerate form of discourse (1962, §35). So even though Heidegger recognizes the communicative aspect of language, he denies that every communicative utterance constitutes a discourse. "Idle talk" provides a good example for this deficient mode of discourse.

In *Being and Time*, Heidegger speaks of "idle talk" as a degenerate form of discourse. Idle talk "does not communicate in the mode of a primordial appropriation of . . . being, but communicates by *gossiping* and *passing the word along*" (1996, 168). Whereas the communicative structure of language opens up a world of shared meaning among the interlocutors (we will return to this point shortly), communication in the form of idle talk is nothing more than a proliferation of groundless information. In other words, idle talk reduces communication to a mere instrument that "[passes] the word along." As such, despite serving a communicative purpose, idle talk lacks the disclosing quality that language can offer.

Furthermore, the communication/language distinction is crucial because even though Heidegger denies language to animals, he does not deny communicative capacity to them. For instance, he acknowledges that worker bees can "communicate information" to their "bee community" via a dancing ritual (1995, 186). So minimally the bees are capable of communication—but still a diminished form of communication that doesn't constitute language. Heidegger also makes a distinction between "words" and "noises"—the former are meaningful and constitutive of human discourse, while the latter are merely sounds produced by the animals, triggered by particular physiological conditions (1995, 307). For Heidegger, even though the sounds of animals do "indicate something" among the animals (thus they serve a communicative function), they still lack meaning because the "animal does not mean or understand by its call" (1995, 307). I will come back to the question of animal communication later, but for now, suffice it to say that Heidegger sees no conflict between denying animals language and granting them communication.

## "AS-STRUCTURE" IN PROPOSITIONAL STATEMENTS

As noted, Heidegger aims to establish the thesis that "man is world-forming." His discussion on animality (and in particular animal's poverty in world) is ultimately at the service of defending this thesis. Because of this comparative approach, it is not until the last chapter of the lecture course that Heidegger finally examines the "man-as-world-forming" thesis. He begins by reminding us that the manifestedness of beings *as* beings belongs to the world. The world is bound up with this "enigmatic 'as,' beings *as* such . . . 'something *as* something'" (1995, 274). And importantly, this "enigmatic 'as'" is a "possibility . . . quite fundamentally closed to the animal" (1995, 274). So what is this "as" that is "refused to the animal" (1995, 287)? What exactly is the animal being denied?

The preliminary step to make sense of this "enigmatic 'as'" is to look at its specific function in a sentence. According to Heidegger, "as" operates as a structural linkage of two terms such that their relationship is made explicit—"'as' signifies a '*relation*' and that the 'as' is never given independently on its own" (1995, 288; emphasis in original). Within a sentence "as" expresses a relation, it gives a sentence a relational structure. This "as" is not given independently in a sentence because there is always something *as* something. This relational structure is rendered intelligible by the expressiveness of beings—"the 'as' can only begin to function if beings are already given, so that the 'as' then serves to render these beings *explicit* as constituted in such and such a way" (1995, 288). Heidegger gives an example of the expression "a *as* b." In order to understand this expression, the meaning of "a" and "b"—that is, "a" being "b"—must be already given (1995, 288). The "as" in this expression is what makes the being (meaning) of "a" and "b" transparent. In other words, the as-structure is what enables meaning to be expressed in a sentence. In *Being and Time*, Heidegger identifies this expressiveness as the apophantic—"pointing out"—function of a statement.[9] In *FCM*, Heidegger confirms that this "as" is a structure that belongs to propositional statements insofar as a statement "[makes] something open or manifest" (1995, 288).

Given this preliminary interpretation of the "as-structure," it seems that for Heidegger what is refused to the animal is the capacity to make propositional statements. Notably, this capacity to make propositional statements involves the ability to *relate*. The structure of relation, in turn, presupposes a network of meaningful terms. As Heidegger repeatedly reminds us, "as" cannot exist independently, "it is a relation which moves from one term to the other—something as something" (1995, 288, 292). The "as-structure" makes sense only against the backdrop of a web of interrelated terms and meaning. So without the apophantic-as—the relational structure that is constitutive of propositional statements—animals do not have access to the network of meaning.

So is linguistic capacity reducible to the capacity to make propositional statements? As soon as Heidegger articulates the relationality of the as-structure, he qualifies his characterization as "formally correct . . . [but] utterly vacuous" (1995, 292). To say that "as" is a relation doesn't really help us understand its *specific* character because, as Heidegger acknowledges, "and" and "or" in a statement *also* express a relation between two terms (1995, 292–93). Like "as," a conjunction and a disjunction also presuppose the interrelatedness of terms and meaning. Like "as," a conjunction and a disjunction also cannot exist independently; there is always "something *and* something" or "something *or* something"! So even though we can characterize "as" in terms of its relational function in propositional statements, Heidegger is adamant that this formal characterization of the as-structure must not be confused with its essence. For Heidegger, in order to get to the "originary dimension" of the as-structure, we must see that propositional statements merely offer a "point of departure" for our investigation (1995, 301). There is an "originary dimension" of the "as" out of which the as-structure of propositional statements arises (1995, 301). This view is in keeping with *Being and Time*, in which Heidegger makes it clear that propositions are merely a derivative mode of interpretation; specifically, the propositional-as is derivative of the more primordial hermeneutic-as (§33–34). For Heidegger, the formal (and empty) characterization of the as-structure "tells us nothing about the 'as' as such, but merely directs us toward our proper and peculiar task [of uncovering the originary 'as']" (1995, 293).

After giving a relational account of the as-structure *and then* pointing out the utter vacuity of this account, Heidegger devotes the rest of the lecture course to uncovering the "originary" dimension of the as-structure. In the end, Heidegger identifies three crucial moments that belong to world-formation: (1) holding the binding character of things towards oneself, (2) completion, and (3) unveiling of the being of beings (1995, §73–74). I will not rehearse Heidegger's argument for this tripartite structure of *Dasein*'s world-formation. Rather, I want to tell a different story of what makes *logos* possible. This story will deviate from the tripartite account that Heidegger offers even though it is still informed by *FCM* and his other writings on language.

As we have seen, for Heidegger the interrogation of propositional statements merely offers a "point of departure" for understanding the relationality of the as-structure. It is important to take this point seriously and move beyond the relational structure in *propositional* statements—after all, saying that humans are distinguished from animals by virtue of the capacity to make propositional statement is obvious and unsatisfactory. It is by moving beyond the derivative mode of the as-structure that we see more clearly what animals are being denied.

## RELATIONALITY AND *LOGOS*

Heidegger highlights the relational character of propositional statements in his formal account of the as-structure. Certainly, Heidegger does not use "relation" (*Beziehung*) to mean kinship (*Verwandtschaft*). Relation here is understood in terms of the interrelatedness of words. However, is there a more primordial level of relationality out of which the structural relation of propositional statements derives? Specifically, what does this structural relation in propositional statements tell us about our *social* relations with others? Does one's ability to articulate "a as b" say anything about one's capacity to relate? And if animals do not relate to meaning in the linguistic realm, can they still relate to others meaningfully in the social sphere? While these questions seem to hinge on an equivocation of the word "relation," they are in fact motivated by Heidegger's interpretation of σύμβολου (symbol).

Notably, Heidegger's account of σύμβολου takes place within the passage in which he discusses the difference between animal utterance (φωνή) and human discourse (1995, §72). The two are distinguished insofar as human discourse is bounded by agreement (κατά συνθήκην); and the genesis of agreement is found precisely in σύμβολου (1995, 307). So following Aristotle, Heidegger isolates σύμβολου as the key ingredient of human discourse. But he finds the concept of "symbol" (the common translation for σύμβολου) inadequate and unsatisfactory. Instead, he articulates the meaning of σύμβολου as follows:

> Συμβολή means throwing one thing together with another, holding something together with something else, i.e., keeping them alongside one another, joining them to and with one another. σύμβολου therefore means joint, seam, or hinge, in which one thing is not simply brought together with the other, but the two are held to one another, so that they fit one another. (1995, 307)

First of all, σύμβολου connotes a relationship—joining two things together. Second, the togetherness of σύμβολου is not some state of random thrown-togetherness. Rather, it is a kind of togetherness in which one thing *fits* another. At this point, Heidegger offers a striking example to illustrate the "joining together" of σύμβολου:

> Whatever is held together, fits together so that the two parts prove to belong together, is σύμβολου. In the original concrete sense, for example, the two halves of a ring are σύμβολα which *two guest-friends share between them and bequeath to their children*, so that if the latter happen to meet later, they can hold together the halves of the ring to see if they fit, and can thereby recognize one another as belonging together, i.e., *as befriended via their fathers*. (1995, 307–8; emphases mine)

Σύμβολου is about joining or bringing together the parts that belong to each other, like the two halves of a ring that fit together. And the two halves of the ring, in turn, are symbolic of a friendship. It is not an accident that Heidegger uses the friendship example to explain σύμβολου: according to the *Oxford Greek-English Lexicon*, σύμβολου also means contract, covenant, bond, engagements of life, and even marriage contract—σύμβολου conveys the sociality of human lives.[10]

In Heidegger's example, the ring is symbolic of more than one friendship, as the friendship between the two friends is passed on to the next generation through flesh and blood. The two halves of the ring are bequeathed to the children of the two friends so that the children may—many years later—recognize that they are "befriended via their fathers." Importantly, the two children do not *become* friends upon reconnecting the two halves of the token; rather, they recognize that they (already) belong together. As such, the children are thrown into a friendship through the father-son kinship—a friendship that they will recognize only after their halved-rings find each other. Importantly, Heidegger uses the German word "*zusammengehörig*" to speak of the "belonging together" of the two children. The word itself is formed by "*zusammen*" and "*gehören*." "*Zusammen*" means "together," "jointly," "common," and "*gehören*" means "to belong," "to appertain." Now "*gehören*" is derived from "*hören*," meaning "to hear," "to listen," and "to hearken." As we know, "hearing" and "listening" occupy an important place in Heidegger's articulation of *Dasein*'s being. For one thing, "hearing is constitutive for discourse" (1962, 163). And insofar as *Dasein* is the kind of being who can engage in a discourse, listening becomes "Dasein's existential way of Being-open as Being-with for Others" (1962, 163). It is in the "belonging" of "*gehören*" that we find this "hearing," which brings us back to the link between discourse and *Dasein*'s being-with—that is, *logos* and sociality.[11]

It is revealing that Heidegger uses the friendship token to illustrate what σύμβολου means "in the original concrete sense" (1995, 307). Specifically, it points to a link between *logos* and relationality. In the lecture course Heidegger maintains that the meaning of words is grounded in convention and agreement (of a community). The occurrence of agreement (κατά συνθήκην) is the condition of the possibility of discourse, *logos* (1995, §72). The relationship between sociality and language can be further illustrated in Heidegger's discussion of communication in *Being and Time* and *Basic Concepts of Aristotelian Philosophy*. As we will soon see, insofar as we are social beings, we are also the kind of beings whose utterances are meaningful. Sociality and discourse intertwine.[12]

## BEING-WITH AND LANGUAGE

We have seen earlier that communication can take two forms: one is the mere proliferation of information, as exemplified in "idle talk"; the other has a disclosing quality that is proper to discourse. We can now turn to the latter type of communication. In *Being and Time*, Heidegger identifies communication (*Mitteilung*) as a structural component of "assertion" (statement) (1962, §33–34). According to him, communication involves:

> Letting someone see with us what we have pointed out by way of giving it a definite character. Letting someone see with us *shares with* [*teilt . . . mit*] the Other that entity which has been pointed out in its definite character. . . . As something communicated, that which has been put forward in the assertion is something that Others can "*share*" *with* the person making the assertion. (1962, 155; emphases mine)

Assertion (or statement), insofar as it communicates, discloses a *shared* world. Discoursing articulates for *Dasein* the intelligibility of being-in-the-world; and being-with—specifically being-with-one-another—belongs to *Dasein*'s being-in-the-world (1996, 162). In other words, communication makes it explicit that the being of *Dasein* is fundamentally being-with-one-another. Heidegger insists that communication is not a bridge that links one subject to the other, as if the two were initially separated. Rather, communication *presupposes* the being-with of *Dasein*, "In talking, Dasein *ex*presses itself [*spricht sich . . . aus*] not because it has, in the first instance, been encapsulated as something 'internal' over against something outside, but because as Being-in-the-world it is already 'outside' when it understands" (1996, 162).[13] Communication reveals to *Dasein* that it is always already socially constituted. In the same vein, the halved-rings that the two friends bequeathed to their children disclose to the children that they are always already part of a friendship.

In his reading of Aristotle's *Politics*, Heidegger articulates the link between speaking and sociality even more explicitly: "*Speaking is, in itself, communicating*; and, as communication, it is nothing other than κοινωνία [community, society]" (2009, 43). Insofar as *Dasein* is speaking-being, it is also social-being. In fact, to the extent that *logos* makes visible our being-with-one-another, being a speaking animal also means being a political animal:

> If λόγος constitutes the having-there-with-one-another of the world, the determination of being-with-one-another is constituted in it. And the determination of the ζῷον λόγον ἔχον [speaking animal] must then, at the same time, contain within itself the determination of the ζῷον πολιτικόν [political animal] . . . this distinctive being-with-one-another is not something that is brought to human beings, but is rather *the* being-possibility. (2009, 40)

Heidegger maintains that our being-with is "equiprimordial" with our speaking-with (2009, 45). It is "equiprimordial" because, on the one hand, speaking presupposes being-with; our words are meaningful because they are bounded by agreement (κατά συνθήκην). On the other hand, being-with is possible by virtue of our speaking capacity—"It is shown that the κοινωνία [community, society] which forms the household (οἰκία) is only possible on the ground of λέγειν [expression], on the basis of the fact that the being of the human being is speaking with the world—expressing itself, speaking with others" (2009, 35). The relationship between speaking and sociality is understood circularly because they are equally basic, equally primordial, and we cannot "deduce one from the other" (2009, 45). As such, while language discloses our sociality, we must understand language and sociality as mutually constitutive.

At this point, let us return to the discussion on language in *FCM*. We begin to see that the formal relationality that we have seen in propositional statements (exemplified by the as-structure) goes hand in hand with our social relations with others. Just as the "as-structure" of propositional statements makes explicit the relation between two terms, σύμβολου is, "in the original concrete sense," a token that discloses a forgotten friendship. According to Heidegger, the occurrence of the symbol (σύμβολου)—that is, the occurrence of this togetherness—is the condition of the possibility of discourse (1995, 308). Indeed, we are able to employ the "as-relation" in propositional statements only because we are already the kind of creatures who are capable of relating socially.

Given Heidegger's interpretation of σύμβολου, in particular the significance of his "guest-friends" example, the animal's inaccessibility to *logos* is shown in a new light. If our speaking-with and being-with are "equiprimordial," then by denying *logos* to animals, Heidegger also calls into question their capacity for social relation. (What is in question is certainly animals' relationship with *Dasein*, but as we will see, even their kinship with other animals seems precarious.) Just as the as-structure in propositional statements is refused to the animal, σύμβολου is what distinguishes human discourse from animal utterance. And if human discourse is bounded *by agreement* (κατά συνθήκην), then animal utterance is mere sound, it is fragmented and unbounded by covenants. The distinction between human discourse and animal utterance is expressed in the two levels of relationality—*both* the formal as-relation *and* the social relation. I will now turn to Heidegger's account of captivation—the specific way animals relate to their environments.

## ANIMAL'S CAPTIVATION

At the end of §48 of *FCM*, Heidegger's interrogation of animality has led him to the question of how the animal relates. He asks, "What does the animal relate to, and what sort of relationship does it have to whatever it seeks as nourishment, seizes as prey, or attacks as hostile?" (1995, 200). This question is particularly interesting because it seems to suggest that animals have some access to the "as-structure" insofar as they relate to others *as* nourishment, *as* prey, or *as* enemy. But for Heidegger, animal relations—if animals relate at all—are necessarily impoverished and circumscribed.[14] And just as he insists that animal sounds are prompted by particular physiological conditions, Heidegger describes animal relations in terms of physiological and instinctual needs.[15]

Two examples (both on the animal's relation to food) from the lecture course are particularly relevant here. In one instance, Heidegger insists that a beetle relates to a blade of grass as a "beetle-path on which the beetle specifically seeks beetle-nourishment, *and not just any edible matter in general*" (1995, 198; emphases mine). The beetle acknowledges the blade of grass as *its* nourishment, but it does not acknowledge the blade of grass as food as such. (The beetle cannot conceive of the blade of grass as food for another animal.) In another instance, Heidegger also denies that bees can relate to food *as such*. He draws this insight from one of J. von Uexküll's experiments. In it, the bee's abdomen was cut off while it was sucking a bowl of honey. The bee reportedly did not stop sucking even while the honey was running out from its body (1995, 242). For Heidegger, this example is indicative of the bee's inability to recognize honey *as such* (honey *as* presence, honey *as* food). The bee is driven by its instinct to eat, and its honey-sucking behaviour is dictated by the sensor of satiety that begins in the abdomen, and thus the bee does not—and need not—relate to honey *as food*. So while animals may relate to others (as nourishment, as prey, as enemy), this relation is necessarily egoistic, driven by a practical, immediate need.

Unlike the beetle and the bee, humans relate to food *as such*. We do, for example, see a piece of steak on the plate as edible matter *in general*—it's edible not only because we happen to be hungry at the time. It is edible matter not just for the specific diner, but also (potentially) for the diner's family, friends, or even her pets—she may request a doggy bag. Indeed, we can even relate to the steak as something other than food. We can, for instance, see the steak as art,[16] as fashion (à la Lady Gaga), as a representation of factory farm, or as the dead mother of an orphaned calf. But despite these various (non-egoistic) ways of relating to the steak, we may question, in a good Derridean fashion, whether *humans* always relate to food as food. Consider the times when we snack mindlessly or distractedly, or the times when we are actually "captivated" by the food and we just want to keep eating. At the end, perhaps the main reason we don't keep gorging ourselves is that we, unlike the amputated bee in Uexküll's experiment, actually have an abdomen.

The egoistic character of animal relation is buttressed by Heidegger's notion of captivation. Again, drawing on Uexküll's observations, Heidegger attributes "captivation" (*Benommenheit*) to the animal as its specific manner of being. Heidegger describes captivation as "that intermediate state somewhere between consciousness and unconsciousness" (1995, 239). This in-between status is particularly appropriate for the animal, which is situated between the world-forming human and the world-less stone. Heidegger describes the bee as being "taken [*hingenommen*] by its food," and that it was "captivated by the scent and the honey" (1995, 242, 243). Furthermore, this captivation is characterized in terms of the "absorption" of the animal.

> Behaviour as a manner of being in general is only possible on the basis of the animal's *absorption* in itself [*Eingenommenheit in sich*]. We shall describe *the specific way in which the animal remains with itself* . . . this way in which the animal is absorbed in itself, and which makes possible behaviour of any and every kind, as *captivation* [*Benommenheit*]. (1995, 238–39; italics in original)

Heidegger's animal is self-absorbed to the extent that it "remains with itself." As Matthew Calarco puts it, the animal is "captive to its own instincts" (2004, 24). The German original for captivation is "*Benommenheit*," which can also be rendered as "dizziness" or the state of being "stunned," "numbed," or "dazed." Derrida calls "*Benommenheit*" a "mute stupor," linking together animal's captivation and their silence (2008, 19). But "*nehmen*," which is the root of "*benommen*," also conveys a sense of isolation or severance as it can also mean "taken away" or "blocked." But from what are the animals being "blocked"? For Heidegger, animals are severed from the world of meaning and thus severed from all linguistic beings, from *Dasein*. Whereas *Dasein* is open to being-as-such, the animal is imprisoned by its instincts and absorbed in itself. Animals and *Dasein* are separated by an "abyss," as Heidegger puts it repeatedly (cf. 1968, 16; 1993, 230).

Heidegger speaks of an animal's captivation as occurring in and through its instinctual drives, and these (interrelated) instinctual drives are constituted in what he calls a "disinhibiting ring" (1995, 249). For Heidegger, this ring opens up to the animal its instinctual capacities (thus "disinhibiting"). This is within the ring that the animal relates to its environment instinctually, though this limited relationship still does not give the animal access to being as such. Indeed, this disinhibiting ring is a ring that encircles the animal in a way that "[the animal] cannot escape" (1995, 249). The animal is a captive of its own disinhibiting ring. Whereas the ring of the two guest-friends reveals to the children that they "belong together" by way of their own flesh and blood, the ring of the animal becomes the barrier that keeps the animal away from the world of meaning and *Dasein*. In the end, Heidegger's ring is at the service of human kinship exclusively.

In Heidegger's account of animal captivation, we see that the animal is denied kinship with *Dasein*. But what about their kinship with other animals? The translation of "*Benommenheit*" as "captivation" also conjures up the image of animals as our literal captives.[17] Animals are held captive by us in zoos, factory farms, laboratories, and even our homes. Our treatments of captive animals seem to presuppose that animal kinship is either absent or irrelevant. In zoos, babies are often separated from their mothers; in factory farms, calves are taken away from their mothers shortly after birth. Even our pets are separated from their mothers and littermates. Isolation or severance is common practice in our captivation of animals. In fact, it has become such a standard practice that we come to see a solitary existence as a form of dehumanized, animalistic existence. In *Solitary Confinement*, Lisa Guenther analyses solitary confinement in relation to animality. As she points out, prisoner advocates often employ the rhetoric of dehumanization when they condemn disciplinary practices—including solitary confinement. She questions whether the rhetoric of "dehumanization" is adequate for us to challenge the practice of solitary confinement, insofar as this rhetoric continues to rely on—and reinforce—the human-animal hierarchy (2013, chap. 6). Prisoners who were subjected to solitary confinement often compare their condition to that of a captive animal. One observer even describes the behaviour of these prisoners as resembling "caged felines pacing at a zoo" (2013, 135). The lack of external stimuli in solitary confinement penalizes the inmates by stripping them of their relationality. However, as Guenther persuasively argues, the rhetoric of dehumanization is problematic as it denies that animals, like humans, are *also* social, intercorporeal beings. The very censure of solitary confinement as "dehumanizing" reinforces the myth that solitude is proper to the animal, while sociality is proper to the human. The rhetoric of dehumanization reveals that we continue to think of animals as asocial beings devoid of kinship in our social-political discourse. While for Heidegger "captivation" deprives the animal of beings as such (i.e., the realm of the symbolic), our everyday captivation of animals also deprives animals of their relational possibility. Once again, linguistic possibility and social possibility become intertwined. In *Animal Lessons*, Kelly Oliver points out that if language for Heidegger is the "house of being and the home of man," then denying animals language would also "[leave] animals with neither house nor home" (2009, 199). Following her analysis, we may say that Heidegger's animals are strays because they have no relations: if family and friendship are constitutive of a home, then the lack of kinship renders animals homeless.

The connection between language and kinship is further illustrated in Heidegger's *What Is Called Thinking?* There Heidegger claims that apes have no hands; he insists that "only a being who can speak, that is, think, can have hands" (1968, 16). The hand is at the service of language (thinking) insofar as thinking is a handicraft. But if having hands presupposes language, then apes don't have hands precisely because they don't speak. What can the hand do that distinguishes it from a mere "grasping organ"? What makes the hand so special that it is reserved for linguistic beings only?[18]

In "Geschlecht II," Derrida argues that the distinction between the hand and the prehensile organ is "reducible to the assured opposition of *giving* and *taking*" (1987, 175; italics in original). In his view, "this thought of the hand belongs to the essence of the *gift*, of a giving that would give, if this is possible, without taking hold of anything" (1987, 173; italics in original). Derrida's emphasis on "giving" is informed by Heidegger's own articulation of the hand as welcoming, "The hand reaches and extends, receives and welcomes—and not just things: the hand extends itself, and receives its own welcome in the hands of others" (Heidegger 1968, 16). Indeed, when handshaking, the hand grabs but only in order to offer. It is by taking hold of the hand of others that one offers hospitality (welcoming) or good wishes (bidding farewell). The hand is the site at which relationship is inaugurated and celebrated. It is distinguished from other grasping organs by virtue of its ability to offer hospitality and nurture relationships. Whereas in *FCM* the openness to meaning—and the possibility of meaningful relationships—is reserved for those who have access to the "as-structure," here the having of hands—and the possibility of generating relationships—is reserved for speaking beings.

What we have seen is that there is a strong link between sociality and language, relational capacity and linguistic capacity. While sociality/relational capacity is reserved for linguistic beings, instincts govern animal behaviour. Recall that the animal does not have access to the "as such" insofar as its relationship with others is circumscribed by its instinctual drives. The beetle can only relate to a blade of grass *as nourishment* because of its instinctual, immediate need to be nourished. The same goes for the bee and the honey.

There is no meaning to the blade of grass or the honey beyond the animal's practical, immediate need. Being "captive to its own instincts," the animal is cut off from the world of meaning. However, if language were proper to relational, social beings, what would be proper for instinctual beings? The answer is mere communication, or non-linguistic communication. As we have seen, Heidegger sees no contradiction in granting animals communication while denying them language; this is so because mere communication and language serve radically different purposes. Whereas language discloses to *Dasein* its being-in-the-world, specifically its being-with-others, mere communication is at the service of instincts or necessities.

Heidegger is not the first (or the only) philosopher who characterizes animal behaviour as "fixed" or instinctual. As Derrida reminds us, the idea that the animal is capable only of programmatic behaviour or a coded message can be traced all the way back to Descartes's notion of the "animal-machine" (2008, 87, 120). Similarly, the idea that mere (non-linguistic) communication is at the service of instincts/necessities is not a novel idea in the history of philosophy.

One particularly elucidating moment can be found in Jean-Jacques Rousseau's *Essay On the Origin of Language* (1986). In the *Essay*, Rousseau gives us an entertaining account of how conventional language (i.e., speech) is invented.[19] According to him, what gave rise to language is our evolution from creatures of mere *needs* to creatures of *passions*. In chapter 2, titled "That the Invention of Speech Is Due Not to Need but Passion," Rousseau argues that it is not possible for speech to be invented for the purpose of expressing our needs. For one thing, we already had gestures to express and address basic needs. For another, the very first need of the human species is to flee from each other and spread across the earth so that we won't overcrowd one corner of the world, so "it would be absurd to suppose that the means of uniting them [speech] derived from the cause of their separation [need]" (11–12). Because conventional language could not have been born out of *needs*, Rousseau posits *passions* as the sources of speech: "it is neither hunger nor thirst, but love, hatred, pity, anger, which drew from them the first words" (12).

For Rousseau, gestures were sufficient for pre-linguistic humans to communicate their needs.[20] He muses that if we ever only have physical needs, we may never have invented speech (1986, 9). Passions, however, cannot be expressed via gestures in silence: "one can take nourishment without speaking. One stalks in silence the prey on which one would feast. But for moving a young heart, or repelling an unjust aggressor, nature dictates accents, cries, lamentations. There we have the invention of the most ancient words" (1986, 12). Speech is invented because we cannot adequately express ourselves by means of gestures alone.

In chapter 9 of the *Essay*, "Formation of the Southern Languages," Rousseau offers yet *another* origin story—this time he explains what gave rise to conventional languages in Southern countries with mild climates.[21] In this origin story, a girl and a boy fell in love after their repeated encounters at a well, where they took water for their family and their herds (1986, 44). According to Rousseau, "[little] by little they become less shy with each other. In trying to make oneself understood, one learns to explain oneself" (1986, 45). Motivated by their passions, conventional language is engendered (1986, 45).

According to this love story, the origin of conventional language cannot be concurrent to the origin of humanity—the boy-meets-girl scene takes place in the context of fetching water for their families. Acknowledging that there was a time before conventional language, Rousseau remarks:

> There were families, but there were no nations. There were domestic languages, but there were no popular ones. There were marriages but there was no love at all. [That is, there was sex but no love.] Each family was self-sufficient and perpetuated itself exclusively by inbreeding. Children of the same parents grew up together and gradually they found ways of expressing themselves to each other . . . natural inclination sufficed to unite them. *Instinct held the place of passion.* (1986, 45; italics mine)

While Rousseau speaks of "needs" in chapter 2 and "instincts" in chapter 9, both are pitted against passions. Whereas conventional language is at the service of love and passions, gestures (pre-linguistic communication) are at the service of instincts/needs. Importantly, Rousseau thinks that families are formed out of need rather than love. It is only when we move from family to society (nation) that we become proper social beings. We go from dwelling in houses to homes.

There are some significant parallels between Heidegger's and Rousseau's views on instinct and non-linguistic communication. For both thinkers, instinctual needs can be adequately expressed and addressed via non-linguistic communication—it is only when we go beyond our instinctual needs that we acquire language. Just as Heidegger's bee flies back to its hive out of its instinct to return (1995, 243), Rousseau's family was united because of a "natural inclination" (1986, 45). Furthermore, both Heidegger and Rousseau connect the lack of language with isolation. Without language, Heidegger's animals were "captivated" and "self-absorbed," while Rousseau's archaic family was "self-sufficient" to the extent that it multiplied via inbreeding (just as many domesticated animals would when kept in an enclosure). For both philosophers, instincts govern behaviour prior to the acquisition of language. For Heidegger, it is the ability to relate to meaning—and the ability to relate at all—that distinguishes humans from instinctually driven animals. For Rousseau, it is passion that gives rise to conventional language and nations; because a nation, unlike a family, is bonded by *love* instead of sexual instinct. While Heidegger speaks of "being-with-others" and Rousseau talks about "passion," for both of them sociality is at the heart of language.

## ANIMAL'S POVERTY IN RELATIONALITY

At this juncture, I should turn to one of the main theses that Heidegger intends to advance in *FCM*—the animal's "poverty in world." There are many interpretations regarding this thesis, especially regarding Heidegger's anthropocentrism.[22] I will not engage in this debate in this chapter; rather, I will examine this poverty-thesis in order to trace another important moment in his notion of animal kinship. As we will see, in Heidegger's illustrations of "poverty" and "deprivation," animal kinship once again becomes his point of departure.

Let us return to the question that Heidegger broaches at the end of §48: "What does the animal relate to, and what sort of relationship does it have to whatever it seeks as nourishment, seizes as prey, or attacks as hostile?" As we have seen, even though this question seems to suggest that the animal has access to the "as-structure" and relationality, Heidegger ultimately insists on a rather impoverished and egoistic picture of animal kinship. The question of how the animal relates to others becomes the focus of §49 and §50. But before he plunges into the question of animal kinship, Heidegger specifies a methodological problem:

> But how are we to do so [inquiring about the animal's relation to the world]? How else than by transposing ourselves into the animal? But do we not then run the danger of interpreting the being of the animal from our own perspective? Perhaps we could ultimately obviate any misinterpretation that might arise. However, there is another much more important and quite fundamental question here: *Can we transpose ourselves into an animal at all*? (1995, 201; italics in original)

Heidegger then extends the question of transposability to stones and humans. Can we transpose ourselves into a stone? Can we transpose ourselves into another human being? Heidegger insists that transposition is not a matter of transference or substitution. We don't, so to speak, put ourselves in the shoes of the other (either actually or imaginatively), as if we can vacate our subjectivity and assume the interiority of other beings. For Heidegger, transposition is not a matter of *becoming* the other; rather, it is a matter of *going along with* the other. The very idea of going along with someone presupposes that there is, in fact, some *other* person with whom we go along. Indeed, transposition is possible only when we maintain a boundary between the other and ourselves. As Heidegger puts it, "There can be no going-along-with if the one who wishes . . . to go along with the other relinquishes himself in advance" (1995, 203).

Significantly, for Heidegger it is the *first* question—our transposability into the animal—that makes the most sense. Or as he puts it in a double-negative way, this question "does not represent an intrinsically nonsensical undertaking" (1995, 204). With regard to our transposability to a stone, Heidegger is adamant that the stone "offers no sphere . . . such that we could transpose ourselves into the stone" (1995, 204). We have no access to the stone not because we lack the appropriate means, but rather because the stone is not accessible to us *in principle*. On the other hand, the question regarding our transposability into other human beings betrays a problematic Cartesian assumption. Specifically, it assumes that each of us begins as a solipsistic consciousness whose being is separated from others. But for Heidegger, insofar as we exist, we are *already* transposed into other human beings. In fact, it is "the essential constitution of human Dasein that it intrinsically means being with others" (1995, 206). Thus it is "superfluous" to ask whether it is possible to transpose oneself into other human beings when this possibility "already and originally belongs to man's own essence" (1995, 207, 205).

Now the question regarding the possibility for the human to transpose into the animal is not "nonsensical." For unlike the stone, the animal does offer the possibility to be transposed. But neither is this question "superfluous," for even though we are in principle capable of transposing into the animal, we are not factically transposed into them. Here enters Heidegger's example of domestic pets. Heidegger insists that a dog may *live* with us in the house but not *exist* with us. The dog may "move within our world" (lying under the table, running up the stairs, eating with us), and yet it *refuses* to go along with us (1995, 210). The dog offers the possibility to be transposed and yet refuses to be transposed. For Heidegger, it is the double movement of offering-yet-refusing that characterizes animal poverty. To see this, it is important to clarify what Heidegger means by "poverty"/being poor.

For Heidegger, poverty does not mean having less or being deficient (vis-à-vis having more or being rich); poverty is not a measure of quantitative differences (1995, 195). This is crucial because Heidegger insists that the talk of the poor-in-world animal and the world-forming *Dasein* should not be understood as a "hierarchical evaluation" (1995, 194).[23] Rather, "poverty" means being *deprived* of something; it is a "lacking or absence of something which *could be present* and generally ought to be present" (1995, 195; italics mine). One can be *deprived* only if one is able to have in the first place. Consider, for example, taking away a toy from a child in contrast to picking up a toy from a chair. We may say that the child is temporarily *deprived* of her toy, but the same cannot be said about the chair even though in both instances a toy has been removed. As such, poverty (understood as deprivation) presupposes the possibility to *have* something in the first place. As Heidegger puts it, "not-having *in* being able to have is precisely *deprivation*, is *poverty*" (1995, 211).

The poverty thesis has often been read in relation to the animal's lack of language.[24] This reading is accurate to the extent that Heidegger anchors his argument of world-hood in the linguistic divide. But let us return one more time to the dog that refuses us. The domestic pet example is supposed to illustrate the notion of "privation" or "poverty." The dog's refusal *presupposes* the possibility for transposition in the first place. (After all, it doesn't make sense for the dog to *refuse* something that it couldn't have in the first place.) Given its possibility for transposal, the dog's refusal to be transposed has rendered it deprived of transposition. Indeed, the dog is deprived of, first and foremost, a relationship with people, even though they live under the same roof. It lives in our physical sphere (our house) without entering the social sphere (our home). The dog is deprived of relationality not because it cannot relate in principle, but because it is caught in a social limbo: it refuses relation despite offering this possibility. As such, the question regarding our transposability into the animal is not superfluous because relationality is not something that we can take for granted in the animal realm. Whereas *Dasein*'s sociality is a given, animal kinship is precarious.

Significantly, the word "refuse" in the original German text is "*versagen*." The root of "*versagen*" is "*sagen*," meaning to speak or to say. The word formation "*ver-sagen*" is especially telling. In German the prefix "*ver*" sometimes changes the meaning of the verb to its reversal; for example, "*achten*" means "to esteem" or "to respect," but "*verachten*" becomes "to despise," "to condemn"; "*lernen*" means "to learn," but "*verlernen*" means "to forget." Sometimes "*ver*" changes the verb to mean that the act is being done in a wrong or undesirable way. For example, "*laufen*" means "to move" or "to walk," but "*verlaufen*" means "to stray"; "*sprechen*" means "to speak," but "(*sich*) *versprechen*" means "to make a slip of the tongue." Sometimes "*ver*" also carries the implication of excess or the progress to destruction. For example, "*bluten*" means "to bleed," but "*verbluten*" becomes haemorrhage or even bleeding to death, and whereas "*schlafen*" means "to sleep," "*verschlafen*" becomes "to *over*sleep."

In light of these modificatory effects of "*ver*," how should we make sense of "*versagen*," and especially *animals'* "*versagen*"? Minimally, their "refusal" has something to do with speaking (*sagen*), and the prefix "*ver*" seems to change this speaking or saying into something negative. (In addition to the verb "refuse," "*versagen*" also means to fail or to malfunction.) So is animals' refusal a failure to speak? Is it a speaking that has gone astray? Or is "refusal" a contaminated, degenerated way of speaking? Whether Heidegger intends it or not, the refusal of animals brings us back to the question of language, and specifically the linguistic divide between *Dasein* and animals.

## LANGUAGE ACQUISITION AND SOCIALITY

I have made a connection between our relational capacity and our language capacity in Heidegger's works, primarily in *FCM*. While I have identified the connection by doing textual analysis, a relational account of language can be supported by studies in infant development and observations of non-human primates. Child psychiatrist Stanley Greenspan and philosopher Stuart Shanker's 2004 work *The First Idea* is particularly illuminating. *The First Idea* is committed to the notion of evolution continuity: qualities that we consider uniquely human (such as language) can be traced all the way back to our prehuman ancestors, and we can learn how we acquire these qualities by observing our close primate cousins. With their endorsement of evolution continuity, Greenspan and Shanker may appear to be unlikely allies with Heidegger—after all, Heidegger insists that the human-animal divide is an "abyss." Nonetheless, there are surprising parallels between Heidegger's and these contemporary accounts of language.

In *The First Idea*, Greenspan and Shanker propose that early emotional interactions with caretakers are crucial to the development of symbols and language. Their hypothesis goes against a typical assumption in the psychology of language, the assumption that emotional and linguistic developments are two distinct and independent processes. For Greenspan and Shanker, not only are the two intimately connected, but "emotions [are] at the very heart of the language development process" (2004, 210). They believe that emotional signalling is essential for symbol creation, and symbol creation, in turn, is essential for language formation.

Greenspan and Shanker identify two necessary conditions for the creation of symbols. First, an image must be invested with emotion in order to become a meaningful symbol (2004, 25). For example, "kitty" conjures the image of a four-legged orange animal with a long tail, but this image must have *emotional* significance to become a symbol—"kitty" means a warm and loving buddy who cuddles at night (2004, 25). The second condition is that perception must be separated from action (2004, 25). For most animals and human babies, perception triggers action: "A trout, for example, perceives a fly or a large fish, and what happens next? He moves toward it and attacks, or he flees" (2004, 26). Likewise, a baby perceives its mother, which triggers hunger and crying. Typically, a perception is tied to certain "fixed" actions or behaviours. It is only when the baby learns how to perceive without taking action that the perception can transform into a "freestanding image." When an image becomes "freestanding," the baby can invest emotional meaning to it and turn it into a *symbol*. In the case of the crying baby, when the perception of the mother no longer triggers hunger and crying, the baby can now associate the mother with other emotional experiences—a mother can now *symbolize* love, for example (2004, 27).

Importantly, the baby learns how to separate perception and action by engaging in what Greenspan and Shanker call "emotional signalling," a back-and-forth emotional exchange between the baby and the caregiver. These emotional exchanges must be "reciprocal" and "co-regulated": both the baby and the caregiver must initiate and participate (reciprocal) and their exchanges should have an influence on each other's behaviours (co-regulated). Indeed, what Greenspan and Shanker describe is a *social* interaction: the baby learns how to properly socialize with its caregiver in emotional signalling. Just as social norms govern behaviours, the baby and the caregiver co-regulate each other's behaviour. As such, social skills are cultivated and honed in the very process of emotional signalling. According to Greenspan and Shanker, in most families these social interactions are routinely carried out. In fact, it is so routine that most of the time we do not register that these interactions are taking place. However, "virtually all children who develop language problems" are deprived of such social interactions—either for a biological reason (children with autism) or environmental reason (caregivers fail to reciprocate) (2004, 197).[25]

Greenspan and Shanker's account of language development offers two important points of comparison with my reading of Heidegger. First, both accounts suggest that a symbol becomes *meaningful* when the meaning is detached from one's immediate and basic needs. Recall that for Heidegger, animals do not have access to the "as such": the meaning of a blade of grass for a beetle is circumscribed by its instinctual need to nourish. Unlike a beetle, however, *Dasein* can invest various meanings to food that goes beyond their need to satisfy hunger. Similarly, in Greenspan and Shanker's account, action-triggering perception is different from a freestanding image. When perception is tied to action, the baby is typically triggered to act in order to satisfy an immediate, basic need (seeing mother = wanting milk); when an image is freestanding, however, the baby is able to invest different meanings in the image independent of its immediate need. In fact, food is often invested with emotions—a peanut butter and jelly sandwich, despite its practical function to satisfy hunger, may represent a carefree, loving childhood. Comfort food and food that we consume during the holidays are often symbolic. (Few people bake a whole turkey at home *just* to satisfy hunger, for example.)

Second, in both accounts sociality is pivotal to linguistic capacity. For Heidegger, language discloses the social world to which we have always belonged. Insofar as "speaking with" is equiprimordial with our "being-with," our capacity to speak goes hand in hand with our capacity to relate to others. Similarly, in Greenspan and Shanker's account, the acquisition of language is grounded in the baby's ability to negotiate emotions with its caregiver. Without these reciprocal and co-regulated interactions, a baby has problems detaching action from perception, which ultimately hinders language development.

This chapter is a response to the link between reason and language that has pervaded the history of Western philosophy. Drawing on Heidegger's work on animality, I articulated an alternative way to conceive of language. Specifically, I showed that for Heidegger, relationality, rather than reason, is at the heart of language. We are the kind of beings who speak not because we are rational, but because we are fundamentally *social* beings. Our ability to relate to the realm of meaning goes hand in hand with our ability to relate to others meaningfully. This alternative account of language, in turn, casts a new light on the human-animal linguistic divide—by denying language to the animal, we are also denying them the capacity to relate. In the end, the "poverty" of animals comes down to a deprivation of kinship.

## TRANSPOSABILITY AND SPEAKING FOR ANIMALS

In closing, I want to turn to the question of speaking for others and particularly speaking for animals. Heidegger's notion of transposition is instructive to the problem of speaking for others. As we have seen, Heidegger insists that transposition means "going along with" the other rather than assuming the position of the other. He is clear that the "otherness" of the other must be preserved in the process of transposition—"the other being is precisely supposed to remain *what* it is and *how* it is" (1995, 202). Indeed, there is no one to go along *with* if there is no otherness.

In her recent book *Entangled Empathy*, Lori Gruen also cautions us that empathizing with others does not mean that we lose ourselves in the perspective of the other. She writes: "The empathizer is always attentive to both similarities and differences between herself and her situation and that of the fellow creature with whom she is empathizing" (2015, chap. 3). On the one hand, similarities between the other and ourselves remind us that we are always already "entangled" in some relationship. On the other hand, the difference between the other and us is what enables us to *have* a relationship in the first place, for it "allows us to preserve the sense that we are in relationship and not merged into the same perspective" (2015, chap. 3). Once again, our differences with the other are precisely what make relationships possible.

How can we sustain the "otherness" in transposition? Is it by letting animals "speak for themselves"? I argued in chapter 1 that even the rhetoric of having an animal speak for itself upholds rather than challenges the authority of speech. As such, the very refusal to speak for the animal continues to assert the primacy of *logos* and linguistic beings. So perhaps a more productive way to sustain this "otherness" is *not* to insist that the animal also speaks, but to acknowledge the asymmetry of our moral responsibility—we are responsible to animals even if they do not reciprocate. For animal ethicists, the fact that animals are "cruel" to each other and that they consume each other does not exonerate our cruelty against animals. Their supposed lack of moral agency is irrelevant when considering our responsibility to them.[26] And an implication of this asymmetry of responsibility is human exceptionalism: we *can* be compassionate even to the most vicious predator, and we *can* adhere to the strictest vegan diet while allowing our cats to chase, catch, and terrorize other creatures in our backyards. We hold ourselves to a different—exceptional—moral standard because we acknowledge our otherness from animals. In the end, perhaps this is what transposing into the animal comes down to: we must take on the responsibility to speak for them even if this responsibility carries an anthropocentric and distasteful history. We must go along with the animal even if they continue to refuse us. It is when we speak for them that we most effectively "go along with"—and mitigate violence against—the animal.

## NOTES

1. Elsewhere, Heidegger insists that *logos* understood as reason (*ratio*) has hindered the "proper unfolding of the question of the world" in the history of western metaphysics (1995, 290).
2. He makes a similar point in *Being and Time* (1996, 165). Page numbers refer to the German page for all *Being and Time* references.
3. The 1929/1930 lecture course contains Heidegger's most extensive account of animality. Like many commentators, my analysis will focus primarily on the lecture course. However, it is important to note that some commentators have employed other texts of Heidegger's to envision a Heideggerian animal ethics that goes beyond the apparent anthropocentrism in the lecture course. For instance, Cave (1982) extends the Heideggerian notion of "Care" to animals; Foltz (1993) advocates a Heideggerian "letting be"—an "ontological humility" in animal ethics (85); and Zimmerman (1983) argues that Heidegger offers a non-anthropocentric way of dealing with our environmental crisis.
4. For example, Colony (2007) argues that Agamben's comparison of human boredom and animal captivation is ultimately flawed because he fails to account for the central role of temporality in human boredom. Winkler (2007) argues that Heidegger acknowledges temporality and historicity in *all physical lives*, thus the return to the element of time in his lecture course actually allows him to "attenuate in a way the abyss between man and animal" (536). Aho (2007) contends that Heidegger was more interested in the "temporal 'event' (*Ereignis*) that 'gives' (*gibt*) meaning to beings" than ascertaining the metaphysical distinction between humans and animals (110–11).
5. Aho (2007).
6. See, for example, Derrida (1987); Calarco (2005); Glendinning (1996); Oliver (2009).
7. In his 2006 essay, Stuart Elden offers a helpful analysis on Heidegger and animals, focusing primarily on animals' relation to *logos* and calculation.
8. See Heidegger (1962).
9. Heidegger (1962, §33).
10. Liddell and Scott (1996, 1676).

11. One of the derivations of "*zusammengehörig*" is "*Zusammengehörigkeit*," which means "togetherness," "solidarity," "common bond," even "shared identity." I am indebted to Susanne Bates for her assistance in the translation of "*zusammengehörig*."

12. I am indebted to Elizabeth Jelinek for her assistance in the Greek translation.

13. This passage expresses a similar sentiment: "[communication] is never anything like a conveying of experiences, such as opinions or wishes, from the interior of one subject into the interior of another. Dasein-with is already essentially manifest in a co-state-of-mind and a co-understanding. In discourse Being-with becomes 'explicitly' *shared*" (1962, 162).

14. As some commentators have noted, animals do not have access to the apophantic-as, even though they may have access to the hermeneutic-as. See Winkler (2007).

15. As we have seen, Heidegger insists that animal sounds must be contradistinguished from words (1995, 307).

16. For instance, Zhang Huan, a contemporary Chinese artist, is famous for his muscle suit made of meat.

17. Winkler offers a helpful etymological analysis of "*Benommenheit*" as follows: "*benommen* (captivated, stunned, taken away, blocked), *eingenommen* (taken in, absorbed), or *Benehmen* (behaviour) . . . all refer back to the German verb *nehmen* (to take), which derives from the Indo-European root **nem*, to distribute, to allot, to assign" (2007, 525). According to Winkler, the significance of "distribution" in "*Benommenheit*" lies in Protagoras's story of nature. In Protagoras's story, Epimetheus was responsible for providing animals with their respective nature, and Epimetheus did so by distributing various capacities to them—a "labour of nature" that Protagoras repeatedly describes with the term "*nomos*," norms (2007, 524). Drawing on Protagoras's story, Winkler argues that Heidegger uses "captivation" to convey the normativity of animal behaviour, thereby challenging the norm-nature binary (2007, 524ff.). While Winkler's provocative analysis focuses on the normativity of animal behaviour, I want to draw attention to the ways animals become our literal captives. See also Derrida's discussion on animal captivation (2008, 19).

18. At first glance, Heidegger seems to be making a distinction between physical and mental grasping. With "grasping organs," one may grab, snatch, or capture, but it is with "hands" that one may craft thoughts and philosophize. But this distinction is complicated by the metaphoric use of "grasping" to mean "understanding." In French, for example, there is an etymological affinity between "taking hold of" (*prendre*) and "understanding" (*comprendre*); whereas in English there is "apprehend," which means on the one hand seizing or arresting (grasping physically), and on the other hand understanding (grasping conceptually). In "Geschlecht II," Derrida argues that the metaphoric use of "grasping" obscures the ontological distinction between a hand and a prehensile organ. In his view, at stake is not a distinction between physical and conceptual grasping, but between "giving and taking."

19. Rousseau makes a distinction between "natural" languages (gestures) and "conventional" language (speech). He is primarily interested in the invention of speech. He thinks that some animals could communicate via gestural languages, but only humans are capable of speaking.

20. The same is true of animals, according to Rousseau (1986, chap. 1, 14).

21. According to Rousseau, this origin story in chapter 2 is applicable to primitive languages; however, he still has to account for why there are *different* languages (1986, chap. 8). He believes that these differences are due to the climate of the region in which the language is formed.

22. See Colony (2007, 12); Winkler (2007, 522–23). Both Colony and Winkler agree that reading Heidegger's thesis as straightforwardly anthropocentric is inadequate.

23. As Oliver (2009) points out, a hierarchy would suggest that "humans and animals had enough in common to be compared"—a claim that Heidegger cannot accept because he views humans and animals as *radically* different beings (197).

24. See Winkler (2007, 523).

25. Greenspan and Shanker's account of language development has significant implications. Like language, morality (another purportedly unique human trait) can also trace its origin story to empathy. In *Primates and Philosophers* (2006), de Waal argues that morality is evolutionary continuous. Morality is not a cultural "veneer" imposed upon the "naturally evil" humans; rather, as *social* beings, it is in our nature to be cooperative and empathetic. Accordingly, de

Waal argues that the origin of our moral capacity can also be traced back to our evolutionary ancestors.

26. I am not suggesting that animals cannot grasp moral responsibility in principle. In fact, animal research seems to suggest that some animals can be moral agents (Shapiro 2006).

# *Chapter 4*

# Animal Identity

## *The Problem of Difference in the Animal Rights Discourse*

### SOME ANIMALS ARE MORE EQUAL THAN OTHERS

The first chapter of Peter Singer's seminal work *Animal Liberation* is titled "All Animals Are Equal." Two questions immediately arise from this title: In what sense are all animals equal? And what counts as an animal—is it a biological or ontological designation? Singer devotes a significant portion of his argument to answering the first question (the "equality question"). For Singer, animals are equal insofar as they are sentient beings, and sentient beings are equal in the sense that their interests deserve equal consideration. While Singer painstakingly addresses the "equality question," he pays much less attention to the "identity" question—what (or who) counts as an animal? In fact, throughout his argument the identity of the animal is assumed rather than interrogated. This is not surprising given Singer's utilitarian commitment: the only morally relevant feature of an animal is its capacity to suffer. The important question is not *what* counts as an animal, but *which* animal counts? A non-sentient creature might be taxonomically classified as an animal, but it would not be an animal that *matters* in a utilitarian moral calculus.

Despite the moral aspiration of "all animals are equal," some animals appear to be "more equal" than others, especially when we look at animal protection campaigns on the ground. In an article provocatively titled "Stop Saving These Animals," Eric Andrew-Gee urges animal advocates to carefully consider their priorities. He argues that some animal protection campaigns are getting more attention and momentum not because the animals in question suffer most, but because they are more "photogenic." He cites Russia's recent ban on the import of harp seal pelts as an example of a campaign with a misguided priority as seals are killed on a much smaller scale than other animals, especially animals that we raise for consumption. He illustrates the absurdity of such a misguided priority with the following comparison: "[it] would be like sending humanitarian aid to a small colony of starving supermodels, while ignoring a famine in a country of five million frumpy cleaning ladies" (Andrew-Gee 2012).

Despite the problematic analogy—comparing photogenic animals to "starving supermodels" and not-so-cute animals to "frumpy cleaning ladies"—Andrew-Gee reminds us that "all animals are equal" remains a theoretical aspiration. Some animals are given priority because they are "intelligent" or "human-like" (e.g., primates and dolphins); some are privileged because they are endangered; some seem to count more just by being cute and cuddly. Given our limited resources, it is perhaps necessary to prioritize certain animals in animal advocacy. However, if prioritizing animals should not be based on the "cute factor," what might be a more legitimate criterion? Which animals should we save first? And which ones should we "stop saving"? These questions draw attention to the problem of identity in animal ethics. The familiar problem of essentialism in identity politics resurfaces in animal advocacy—the "animal" in animal ethics is so inclusive that it eclipses the internal differences in the category of what we call "animal." After all, who (what) are we speaking for in animal ethics?

In this chapter, I analyse issues of identity within the animal protection movement. I will begin by examining the messiness of animal identity, a messiness that has often been concealed under the seemingly inclusive category of "animal." The purported inclusivity of animal ethics obscures our dissonant—and at times conflicting—commitments to different animals. The question of animal identity is also intimately tied to the question of language, specifically the question of how we should speak for animals. Borrowing insights from feminist literature on identity politics, I will articulate the strategies of identification and counter-identification employed by animal advocates.

While Jacques Derrida's critique of essentialism motivates my analysis of identity troubles in animal advocacy, Ludwig Wittgenstein's notion of family resemblance helps me conceptualize the human-animal relationship as a nonlinear, crisscrossing web of families. Although Derrida and Wittgenstein belong to different philosophical traditions, they are both resistant to the essentialism that has plagued our understandings of concepts, definitions, and identities. Engaging the two thinkers will allow us to reconsider the "problem of difference" in identity politics.

Although this chapter is motivated by questions of animal identity within the animal protection movement, examining the concept of the "animal" brings us back to some key issues concerning language: the violence of speaking for others, the special power of silence, and the inevitability of translation. In the last sections of this chapter I will return to issues of violence, silence, and translation by examining the significance of "*animot*," a word coined by Derrida. This neologism signifies Derrida's effort to resist the homogenizing presupposition of the term "animal." By examining translation issues surrounding the word "*animot*," I reinforce my argument in chapter 1 that there is no such thing as a free-floating, autonomous animal voice free from human representation. Insofar as we are speaking beings with a political voice, we cannot abnegate our obligation to represent animals, even at the risk of projecting and imposing our desires onto their voice.

## ANIMAL IDENTITY IN THE ANIMAL PROTECTION MOVEMENT

Our failure to register various identity troubles in animal ethics is due in part to how the "animal" is typically invoked. Within the history of Western philosophy, the "animal" often serves as a foil against which we acquire self-identity—we are what the animal is *not*. Thus the question of animal identity is often at the service of constructing human identity. This negative anthropology, however, presupposes a unified animal identity. This unity is achieved by putting *all* animals in a dichotomous relationship to humans—the animal is that which is *not* human. Thus we are caught in a circularity wherein we define the human as what the animal is *not*, and we define the animal as what is *not* human; neither the human's nor the animal's identity is determined. But the question of animal identity is more than a metaphysical or ontological puzzle. As the example at the beginning of the chapter shows, the animal protection movement is fraught with identity troubles.

The indeterminacy of animal identity is particularly evident in legal discourse. While the label of "animal" is inclusive, animal protection agendas do not (and cannot) include *every* species of creatures that we typically call "animal." In fact, animal laws typically protect only vertebrates; given that vertebrates make up only 2 percent of all animal species, *most* animal species are in fact excluded from animal protection laws. In the United States, each state has its own animal protection laws and the definition of "animal" varies across states. For example, in Alaska "animal" includes all vertebrate living creatures except humans and fish; in Illinois the definition of "animal" also excludes humans, but it is defined more liberally to include "every living creature, domestic or wild." In Colorado, Ohio, and Wyoming, "animal" means "any living dumb creature"—a definition that invokes the linguistic divide between humans and animals.[1]

Even within the same state, the meaning of "animal" may change depending on the offense. For example, Iowa penal code stipulates the meaning of "animal" as non-human vertebrates. When it comes to general prohibitions such as animal abuse and animal torture, this definition of "animal" has a long list of exceptions: it excludes livestock, game, fur-bearing animals, fish, reptiles, amphibians, and any non-game species that are considered a nuisance. However, the definition of "animal" changes when it comes to bestiality. For the purpose of defining bestiality, an "animal" becomes "any nonhuman vertebrate, either dead or alive," and the long list of exceptions disappears. In other words, when it comes to animal abuse and animal torture, Iowa law does not necessarily protect creatures such as livestock and game, and it certainly does not cover dead animals. But when it comes to bestiality, the definition of animal *expands* to include even the dead creatures. That a dead animal cannot be abused or tortured is understandable from the animal welfare perspective, but if it cannot be abused or tortured, can it be sexually assaulted? This definitional discrepancy suggests that even though the prohibition against bestiality is catalogued under the title "Sexual Assault," it goes beyond the concern for animal welfare. It is considered a taboo—punishable by law—to engage in bestiality *even if* the animal in question is already dead and no longer calls for protection.

Generally speaking, the animals that matter within the legal realm are restricted to non-human vertebrates, while "pests" are typically excluded from anticruelty law.[2] But just as our legal discourse privileges certain species of animals, public sentiment towards animal welfare also reveals favouritism. For instance, while dogfighting enrages the public, slaughterhouses, animal circuses, and trophy hunting are generally sanctioned. Someone can be simultaneously a dog advocate and deer hunter and someone who vehemently opposes cock-fighting may not have any qualms about enjoying a chicken sandwich.

Animal advocates have criticized "single-issue" campaigns such as the anti-fur campaign and the anti-dolphin-fishing campaign. Some of these single-issue campaigns are more insidious when they sidestep the more systematic, institutionalized exploitation of animals. In a commentary on West Hollywood's ban on fur, Gary Francione criticizes these campaigns for "[seeking] arbitrarily to declare some form of animal use or some animal product as morally more odious than others. These campaigns are not only problematic as a matter of moral theory; they are a practical failure in real-world terms."[3] For Francione, as long as we remain complacent about the *systematic* use of animals, picking and choosing one issue to protest will not improve animal welfare in general.

Of course, many animal advocates are concerned with the institutional exploitation of animals—there have been campaigns for animals in slaughterhouses and even for the less popular animals such as lab mice and spiders. However, the inclusivity of the label "animal" remains problematic. Different species of animals call for different—and at times competing—obligations; fulfilling a duty to one animal is often made at the expense of other animals. For example, proponents of animal testing often attempt to justify their cause with the claim that animal testing saves *animal* lives as well.[4] This justification appeals to the well-being of animals—but only animals that matter to us. Although animal rights advocates typically oppose animal testing, some animal protection agendas are predicated on this economy of life-exchange. For instance, the preservation of a carnivorous species at conservation centres necessarily involves the death of other animals as prey (as food). Take another example closer to home: adopting a cat means that we are introducing a carnivore to our home. Our duty to our cat must be fulfilled at the expense of other animals. Our concern for the well-being of one animal often comes at the expense of other animals.

Even if we could eradicate the institution of animal exploitation, we cannot end animal suffering. While we may protect animals from cruelty done by humans, we cannot protect them from suffering caused by other animals. Insofar as death and suffering are unavoidable in nature, the scope of animal protection is circumscribed. After all, we cannot (and should not) stop carnivores hunting their prey. We cannot protect animals *from other animals*. The inclusivity of the name "animal" in animal ethics can turn against itself and obscure our conflicting and dissonant commitments to different animals.

## ANIMAL IDENTITY, ANIMAL STANDPOINT

Questions of identity are intimately tied to the question of speaking for animals. As I argue above, we cannot simultaneously speak for *all* animals. Advocating for chicken and fish, for example, may compromise our obligation to our companion animals.[5] But setting aside our conflicting obligations to different animals, can we responsibly speak for those whose standpoint we do not (and perhaps cannot) occupy? How can we "put ourselves into another's shoes" when this "other" has paws, claws, or talons? This is perhaps why animal advocates often employ the "like-us" strategy: by emphasizing the *shared* qualities between humans and animals—be it sentience, vulnerability, or the capacity to experience life—animal advocates seek common ground on which we can speak for animals. As such, even if we do not know what it is like to be a bat, even if we cannot occupy the standpoint of a bat, we can still speak on its behalf from the vantage point of a sentient being or vulnerable subject. But as we saw in chapter 1, the act of speaking for others, even if well intended, can be problematic. Feminist philosophers have cautioned against the paternalism that often comes with the impulse to speak for others. Nonetheless, the "violence" of speaking for others is more than assuming the identity of others in a paternalistic way. Speaking for others can also be "violent" if we speak for them *as a unified group*, that is, if we fail to respect the heterogeneity of those for whom we speak.

The problems of animal identity are forcefully articulated by Derrida in *The Animal That Therefore I Am*. There Derrida takes issue with the name "animal"; he argues that the name "animal" is employed so that we can "speak of the animal with a single voice" (2008, 32). We use the word "animal" in the singular as if it represents *all* animals—from our primate cousins to our feline companions, from birds to insects to sponges. We reduce all these different species of living thing into an imaginary unity, a "chimera," as Derrida calls it (2008, 23). For Derrida, whenever we use "animal" in the singular we are complicit in a "veritable war of the species" (2008, 31). Perhaps we can even think of it as a war *on* species. In this war, we do violence to animals not with a butcher knife, but by refusing to name them properly. We reduce biodiversity by reducing different species to a homogenous singularity.

Derrida's critique parallels a concern many feminist, race, and queer theorists have considered in identity politics. Specifically, the concern is that labelling a movement under a general category (such as "woman," "gay," "African American") can obscure internal differences within the movement.[6] For instance, the women's movement has been faulted for overlooking racial and class differences among its participants. In *Ain't I a Woman: Black Women and Feminism*, bell hooks argues that the feminist movement has been shaped largely by the vision of middle-class white women, a vision that is either irrelevant or antithetical to the needs of women from different social classes or ethnic backgrounds. So despite its inclusive label, the women's movement is at the service of a very specific class and race of women.

In the animal protection movement, the label of "animal" engenders similar identity problems. As we have seen, not all animals are included in the moral community and even fewer animals are included in animal protection laws. One obvious reason is that some animals simply do not count as they lack features that are morally relevant, such as sentience. They are taxonomically identified as animals, but they lack moral claim—it makes little sense to legislate a law protecting sponges or sea stars from an anticruelty standpoint. However, as the example at the beginning of the chapter shows, some animals are "more equal" than others simply by being cute and cuddly. Just as the women's movement has been at the service of a specific class and race of women, animal protection campaigns have largely been at the service of primates, pets, or endangered species (even though animals raised in factory farms are also sentient). And just as the interest of middle-class white women might be at odds with the interests of women from a different class and race, our moral commitment to primates and mammals may be made at the expense of other species of animals.

In fact, even the *strategy* of the animal rights movement is bound up with identity issues. Once again, we can borrow insights from feminist discourse. Feminists have identified and critiqued the logic of identification and counter-identification in the women's movement, specifically the strategic shift from *women-can-do-it-too* to valorizing the differences between men and women. Neither of these strategies is satisfactory: the logic of identification continues to privilege men by casting masculinity as the norm, while the strategy of counter-identification is predicated on an oppositional thinking that many feminists seek to overturn. As José Medina points out, both strategies render the feminist movement "parasitic on a masculinist ideology" (2003, 665).

In the animal protection movement, we again see the logic of identification and counter-identification. Early discourses on animal rights/welfare were dominated by the strategy of identification. Animal advocates often stressed the *likeness* between humans and animals in order to challenge the hierarchical distinction between the two. Scientific experiments or observations of animals were often cited to show that animals—*like us*—also have the capacity to speak,[7] use tools,[8] or act morally.[9] As such, the "like-us" strategy has been employed to call into question human exceptionalism.

In the past decade, philosophers have begun to challenge this "like-us" strategy in animal advocacy. Feminist thinkers such as Catharine MacKinnon and Kelly Oliver have critiqued this strategy through an anti-patriarchy lens. For instance, Oliver writes, "Just as feminists ask why women have to be like men in order to be equal, we can ask why animals have to be like us to have inherent value. The notion that man is the measure of all things is precisely the kind of thinking that justifies exploiting animals, along with women and the earth, for his purposes" (2009, 30). As Cary Wolfe also points out, the strategy of identification reinstates the very human exceptionalism that animal advocates reject. Responding to the *Great Ape Project*, Wolfe writes,

> The model of rights being invoked here for extension to those who are (symptomatically) "most like us" only ends up reinforcing the very humanism that seems to be the problem in the first place. To put it very telegraphically, great apes possess the capacities that we possess, but in diminished form, so we end up ethically recognizing them not because of their wonder and uniqueness, not because of their difference, but because they are inferior versions of ourselves, in which case the ethical humanism that was the problem from the outset simply gets reinforced and reproduced on another level. Now it's not humans versus great apes, [it's] humans and great apes—the "like us" crowd—versus everyone else. (2003, 192)

For Wolfe, the problem of the logic of identification is this: if animals are worthy of saving only if they are sufficiently like us, then we have failed to appreciate animals *as animals*. However, what does it mean to ethically recognize animals not in spite of, but because of, their uniqueness and difference?

Against this strategy of identification, philosophers have argued for an alternative model of animal ethics—one that isn't predicated on the animal's likeness to us. As we saw in chapter 1, MacKinnon and Laycock have argued for the need to listen to animals in their own voice and even their own silence. But as I argued, in order to appreciate animals in their *own* voice/silence, we need to translate their voice/silence into terms that are intelligible to us; and as long as translation and representation are inevitable, we cannot avoid contaminating the voice that is purportedly the animal's own. Furthermore, if there is anything we could learn from the women's movement with regard to identity politics, we should be wary of the oppositional thinking that is symptomatic of the logic of counter-identification. After all, don't we fall back on the animal-human binary when we insist that animals should have their *own* voice—in contradistinction to the *human* voice?

Despite the familiar strategic movement from identification to counter-identification, there is a limit to the analogy between the feminist movement and the animal protection movement. The very notion of *identity* seems to impose challenges that are unique to the animal protection movement. For one thing, while we can articulate our *own* identities, animals do not participate in discussions on animal identity. Rather, animal identity remains largely a human (and primarily academic) construct. This is not to say, of course, that animals do not negotiate or contest their identities in a non-discursive way. Consider the example of "Nim Chimpsky," a chimp who was raised as a human infant but was later abandoned to a research facility. In the documentary *Project Nim*, we see Nim asserting his place in his caretaker's family by repeatedly challenging the paternal figure in the family. Nim's struggle to become the alpha male is a non-discursive way of negotiating and contesting his place in the family. Nonetheless, there is still something amiss when we engage in a discussion on animal identity without the direct, non-human-mediated input from those whose identity is being debated. In other words, even if we attend to animals' non-discursive ways of expressing their identities, the very fact that animals require our mediation shows that their participation in the discussion is compromised in a significant way. As such, even though feminist discourses on identity politics have been instructive, we need to be wary of drawing any easy parallels between the feminist movement and the animal protection movement regarding the identity question.

Thus far I have delineated some identity troubles in the animal protection movement. My analysis is very much motivated by Derrida's critique of the term "animal" in *The Animal That Therefore I Am*: insofar as we use "animal" as a category that covers all creatures that are non-human, we reduce multiplicities, differences, and diversities to a human-versus-animal binary. Drawing from discourses on identity politics, I argue that neither the strategy of identification nor the strategy of counter-identification is satisfactory: the former does not escape humanism, whereas the latter is predicated on the human-animal binary. I will return to Derrida's critique of the "animal" as a unified category later on. For now, with Derrida's deconstructive analysis of the "animal" in mind, I turn to another philosopher who also made significant contribution to the critique of essentialism—Ludwig Wittgenstein.

## WITTGENSTEIN'S CRITIQUE OF ESSENTIALISM

In his introduction to *Slow Cures and Bad Philosophers*, a collection of essays on bioethics, Carl Elliott confesses that it is an "uphill battle . . . for any book that claims to find useful guidance from Wittgenstein for bioethics and the practice of medicine" (2001, 1). The same could be said about a chapter on Wittgenstein and animal ethics. It is not that animals have never made their way into Wittgenstein's corpus. Quite the contrary, Wittgenstein has dutifully invoked the animals in his writings. Dogs are Wittgenstein's favourite: dogs have no remorse,[10] dogs cannot fake pain,[11] and (my favourite of all canine references) dogs can neither be hypocritical nor sincere.[12] While dogs are clearly Wittgenstein's best friends, let's not forget about the talking lion (whose words we would not understand),[13] the beetle in the box, the duck-rabbit, the goose that has no teeth,[14] and, of course, the rhinoceros that lurks in the room.[15] Despite this impressive menagerie, it remains an "uphill battle" to conceive of a distinctive Wittgensteinian view of animality in general and animal ethics in particular. The difficulty of this "uphill battle" arises in part due to Wittgenstein's rejection of essentialism and his suspicion of theory, and especially the highly codified moral rules that are so prevalent in contemporary animal ethics literature. (His rejection of essentialism makes any unified, singular Wittgensteinian view on the animal suspect.) Another difficulty arises from Wittgenstein's aphoristic writing style; his terse and often cryptic remarks on animals have been interpreted in different—sometimes contradictory—terms. Indeed, Wittgenstein has inspired theorists from *both* sides of the animal protectionist movement;[16] his insights are invoked to advance a wide array of theses regarding the animal.[17] Keeping in mind that it remains an "uphill battle" to articulate a singular, unified Wittgensteinian thesis on the animal, I will discuss his notion of "family resemblance" as it relates to the problem of difference in identity politics. My discussion is very much indebted to José Medina's 2003 essay "Identity Trouble: Disidentification and the Problem of Difference," in which he articulates a Wittgensteinian response to the problem of difference in identity politics.

Wittgenstein begins his *Philosophical Investigations* with a critique of linguistic essentialism. Linguistic essentialism is the idea (or presupposition) that there is a singular, unified account of language. Wittgenstein illustrates the essentialist theory of language with a long quote by Augustine; in it, Augustine presents an account of language in which words function *exclu-*

*sively* as labels of objects. Wittgenstein presents Augustine's view as such: "In this picture of language we find the roots of the following idea: Every word has a meaning. This meaning is correlated with the word. It is the object for which the word stands" (2009, §1). For instance, the word "table" refers to the four-legged wooden furniture on which I place my laptop; the word "cat" signifies the furry, four-legged animal sitting on the mat; the word "coffee" means the hot, brown liquid I consume every morning; and so on. For Wittgenstein, while this conception of language is not wrong, it describes linguistic practices too narrowly. For him, words do label objects, but this is only one of the many ways words operate, and linguistic functions are not reducible to signification. To capture the diversity of linguistic activities, Wittgenstein coins the term "language-games" (2009, §23).

There are a variety of language games: there is certainly the language game of signification in which words are used as labels. But there are also language games that include irony, jokes, or flirtations—games in which words do more than just signify. (In fact, it would be *unfortunate* to interpret a game of flirtation as a game of signification.) It is in the context of defending his choice of the word "game" in "language game" that Wittgenstein presents his famous account of "family resemblance." There Wittgenstein is responding to the charge that he is "taking the easy way out," for he has not yet provided the "essence of a language-game, and hence of language" (2009, §65). That is, he has not yet offered the necessary and sufficient conditions for what we call "language." In response to this charge, Wittgenstein asks us to consider what we call "games" (2009, §66). We call a wide range of activities "games," from chess to video games, from the Olympics to children playing spontaneously made-up games. While one game may share something in common with another game, there is not a singular quality shared by *all* games. For instance, a basketball game and a game of Monopoly are both competitive games, but not all games are competitive (e.g., solitaire); and while many games are governed by well-defined rules or arbitrated by referees, some games are played spontaneously (e.g., two siblings racing on their way to school). How should we characterize the similarities that we find in these different games—similarities that are shared by *some* games but not by *all* games? Wittgenstein proposes that we think of these similarities in terms of "family resemblance" (2009, §67).

If we look at a family portrait, we see certain features shared by some family members and other features shared by other family members. For instance, the sons might have the same hair colour as the father, and the daughters might have green eyes like the mother, but all the children are over six feet tall, unlike their petite parents, among other things. There is no essential feature that *defines* that family, and yet there are overlapping similarities that we see in family members that *show* that they are a family. For Wittgenstein, different linguistic practices are related to each other like family. There are many different language games and they are grouped together under the label of "language" not because they all share a common core, but because they constitute "a complicated network of similarities overlapping and criss-crossing" (2009, §66).

It is almost uncanny that Wittgenstein uses the metaphor of *family* resemblance in his account of language. As we saw previously in chapter 3, kinship occupies an important place in Heidegger's account of language. Recall that Heidegger's interpretation of "symbol" (σύμβολου) hinges on his "guest-friends" example, in which he analogizes the "joining together" of σύμβολου to the joining together of the two friendship tokens that the two friends pass on to their children as family heirlooms. Nonetheless, despite the coincidental reference to family/kinship, Wittgenstein's family metaphor does not aim at illustrating the social agreement on which meaning is found. Rather, he uses this metaphor primarily to contest the essentialism that has plagued our concept of language. He aims to show how different language games can be grouped together under the same category (family). As such, the "family" in family resemblance has little to do with sociality in Wittgenstein's account of language.

Yet Wittgenstein's family metaphor remains suggestive as we can begin thinking about linguistic *capacity* in relational terms. Indeed, how did we come to acquire our social skills and our language skills? Are these two skills related? The idea that language games are structured like familial relations (crisscrossing and overlapping each other) invites us to consider the connection between language games and social skills: just as we play well in certain language games but are awful at others, we feel at ease in some social circles but feel awkward in others. Our ability to navigate different language games is in part informed and shaped by our ability to navigate different social and familial circles. For example, an animal advocate may learn the language game of protest by joining an animal protection campaign; a recently divorced man may relearn the language game of flirtation by joining an online dating service; a closet lesbian may acquire the language game of secret-keeping, disguise, and even self-denial if she were raised in a conservative family that disapproves of homosexuality. In short, the kinds of language games we learn to play are very much informed by the social and familial circles to which we belong. Insofar as our identity as linguistic beings is constituted by the language games that we play, we are linguistic beings because we are, as it were, social beings.

## A FAMILIAL ACCOUNT OF THE HUMAN-ANIMAL RELATIONSHIP

Employing Wittgenstein's notion of family resemblance, Medina develops a "familial view of identity" (2003, 659ff.). He argues that thinking of identity in familial terms enables us to "dissolve" the problem of difference in group identity (i.e., the view that group identity necessarily obscures internal differences). In his view, insofar as our identities crisscross and overlap one another like family relations, identity necessarily *presupposes* differences and heterogeneities. We all belong to multiple families and we all belong to multiple identity groups. Just as being a member of a family does not mean we are identical to our sisters, brothers, and parents, being a member of an identity group does not mean we are homogenous to other group members. Rather, we are able to identify with a group because we can emphasize the common features *against the backdrop of our differences* with other group members. To use Medina's term, we are willing to be "blinded" to the differences *for the purpose* of identifying with a group. Conversely, we are willing to be "blinded" to the similarities for the purpose of counter-identifying with a group.

Central to Medina's argument is the strategy of disidentification, the strategy that "brings both similarities and differences simultaneously to bear on one's identity" (2003, 664). Unlike the strategies of identification and counter-identification, disidentification is "a way of identifying with the members of a family without losing sight of one's differences with them, or a way of counter-identifying with the members of other families while seeing one's similarities with them" (2003, 664). As such, disidentification serves as a crystalizing occasion in which we finally see that despite our similarities, we are different from our family members, and despite our differences, we also share something in common with members of other families (2003, 664). In other words, disidentification elucidates the multiplicity that is inherent in group identity. Importantly, disidentification is not premised on the binary logic wherein identity and difference are viewed as incompatible or opposites. Rather, the strategy of disidentification presupposes that identity and difference are *constitutive* of each other.[18]

Although Medina is primarily concerned with human identities, his argument is instructive for considering animal identities. If we take seriously Medina's point that identity and difference are not opposites, then we can begin to make sense of the plethora of differences—and similarities—within the category we call the "animal." There are indeed *differences* within the animal kingdom. Just as the category of "language" is constituted by different linguistic practices, the category of "animal" is made up by a network of overlapping and crisscrossing subgroups. (We can think of each of these subgroups as a family unit.) The most typical way to catalogue these subgroups is to divide them according to species membership. However, classifying animals according to their species is only one way of doing taxonomic analysis. Once we move beyond species and begin to consider different categories, it becomes clear that animals can belong to *multiple* overlapping and crisscrossing subgroups/families. There are many different ways to define a family: there can be a family of deceitful animals, a family of playful animals, a family of cannibalistic animals, and so on. A swan and a gibbon clearly belong to different species and inhabit different environments, yet they both belong to the family of monogamous animals. A bat, a dragonfly, and a jellyfish all belong to the family of carnivores despite their obvious differences. In other words, once we consider family units beyond the typical species tags, we begin to see that seemingly disparate animals can be brought together as a group, as a family. The crisscrossing, overlapping feature highlights the heterogeneity of group members. Assigning a swan and a gibbon to the identity group of "monogamous animal" is not to say that they are homogenous; rather, it means that we focus on one particular feature that they share in common (monogamous behaviour) *against the backdrop* of their differences. As such, differences are *inherent* in each identity group.

Given the "family" metaphor, we can even consider *human* relationships with animals in familial terms. Some animals are our more immediate families, whereas some are our more distant cousins. Some animals may be so distant from us that we don't even recognize them as part of our family. It is easy to imagine being kin to our closest DNA relatives—chimpanzees and other primates. But bees and ants can also be our kin if we are looking at a family of creatures with structured social hierarchy. Some of us belong to the family of herbivores by adopting vegetarianism, while some of us share a family with bats and whales by developing the ability to echolocate. The overlapping and crisscrossing character of family resemblance tells us that we may be kin with some unexpected creatures. Insofar as our extended family is connected to us via a network of "immediate" relationships, we cannot rigidly demarcate our family and we must always be ready to accept new family members.

The idea that differences are inherent in each identity group is especially helpful for us to speak about the qualities that we consider uniquely human; for instance, language and moral agency. Does the familial view allow us to talk about human uniqueness without elevating ourselves above the animal? Can we still account for the linguistic capacity and moral agency that typify human uniqueness? I do not deny that proto-language and fledging moral agency can be found in certain animals, but the ways humans have developed as linguistic and moral beings are very different. So is there still a sharp human-animal divide—an abyss between humans and animals, as Heidegger insists? Or do humans and animals exist on a continuum, and whatever differences between humans and animals are merely differences of degree?

If we take Wittgenstein's familial view seriously, we would see that this is a false dilemma. Indeed, the family metaphor helps us articulate our differences from animals without presupposing a sharp and decisive distinction. Just as we can relate to our estranged relatives in different ways (we may feel indifferent towards one and hostile towards another), there is no singular quality that distinguishes us from animals *in general*. The ways we are different from a bonobo are not the same as the ways we are different from a bat—there are different kinds of difference. Just as our linguistic capacity and moral agency set us apart from many animals, the capacity to fly or the ability to sense a tsunami also sets other animals apart from us. There are many differences between humans and animals, but to elevate just one of these differences as *the* difference that separates humans and animals is arbitrary at best and self-serving at worst. As such, acknowledging the differences between humans and animals (be it language or moral agency) need not commit us to drawing a sharp and singular human-animal distinction.

The family metaphor also addresses the second horn of the dilemma, which is the idea that humans and animals are on a continuum and that their differences are merely differences in degree. The family metaphor resists precisely this sort of linear continuity; it recognizes differences that are not easily reconcilable. The fact that we have multiple identities means that we belong to multiple families simultaneously. But just as our identities may generate conflicting commitments, our "families" may not always get along. Consider, for example, a pro-choice Catholic nun. She belongs to both the family of Catholic nuns as well as the family of pro-choice advocates—it would be an understatement to claim that these two families merely differ *in degree*. Yet these two families overlap and crisscross by virtue of the members that they share. As such, we need to reconceptualize the human/animal distinction not as a linear continuum, as if we could have a neat story of amoebas on the one end and humans on the other. Rather, we should look beyond the species tag and appreciate the multiplicity of our differences.

At this point, it is important to evaluate the merits, as well as the limits, of the family metaphor. Articulating our relationship with animals in familial terms opens up a different way of thinking about our responsibilities to them. Specifically, it allows us to speak of obligations in relational terms. It allows us to speak of our responsibilities to animals in terms of our kinship with them and not simply their rights against ours. This is not to say that animals should not have rights or that they are merely objects of care. Rather, just as care and entitlements are both important for humans, the same holds true for animals. After all, we feel obligated to our family members not because they can lay claims on us, but because we *care* about our family.[19] At the same time, we also need rights to protect the vulnerable when there is no one to care for them. As mentioned above, the overlapping and crisscrossing feature of the family metaphor makes it difficult for us to demarcate our family in absolute terms, thus complicating the issue of to whom we respond and for whom we are responsible. Furthermore, the nearly limitless ways we can define a family also suggests that our responsibility to other animals is infinite, akin to a Derridian hyperbolic ethics—an unconditional, impossible, more-Kantian-than-Kant ethics. Such hyperbolic ethics is a reminder that we must be vigilant of our obligations to other creatures. That is, we can never say, *with certainty*, that we are not responsible to certain creatures because they are not our family. Insofar as the boundary of a family is porous and uncertain, our familial responsibility is also undecided and open ended.

Despite its appeal, the family metaphor is not without its drawbacks. First, despite its fluid and malleable boundary, this metaphor remains reliant on the logic of exclusion and inclusion. Even though we may belong to multiple families, we are necessarily excluded from multiple (indeed infinite) families. Insofar as the *marking* of a family involves identifying non-family members, the metaphor of family does not escape the exclusionary logic that has hitherto defined our relationship with animals.

Second, the family metaphor can also be inadequate as it is not immune to violence. As we know, sometimes a family breeds and harbours insidious and unspeakable violence. Victims of domestic violence are often the least willing to speak up against their aggressor; when violence does happen in a family, it often goes unpunished.[20] However, perhaps we should not be surprised that family is not immune to violence. Family dynamics are fraught with complexity as well as contradictions. We may simultaneously love and resent our family. We may continue to love even those who prey on our love—even the ugly, dangerous, or predatory family member who continues to command our care. In fact, this violence may even speak the truth of the animal family dynamic. As long as carnivores feast on their prey, as long as a cat's favourite pastime involves terrorizing squirrels, we cannot eradicate violence in the family of animals. As such, while the family metaphor opens up a space for us to speak of obligations in relational terms, it does not allow us to escape violence. But if family is not immune to violence, what does it mean to have a philosophy of language that gives primacy to kinship and relationality? Perhaps this suggests that language is also not immune to violence, especially when we consider the violence of translation and the violence of speaking for others. In the end, perhaps the worst violence is the fantasy that we, as living beings, could live without violence. And accordingly, the worst violence in language is the fantasy that we could play, as linguistic beings, a cruelty-free language game.

## THE VIOLENCE OF LITERAL AND METAPHORIC CONFINEMENTS

The inescapable violence in the animal family brings us back to Derrida's critique of the name "animal" in *The Animal That Therefore I Am*. As we have seen, Derrida argues that every time we use "animal" in the singular we are complicit in a "veritable war of the species" (2008, 31). Thus far I have focused on the problem of difference, but it is time to examine the violence implicated by the name "animal." In Derrida's reading, the danger of the name "animal" in the singular goes beyond its homogenizing effect, as the category of "the animal" enables us to "corral a large number of living beings within a single concept" (2008, 32). Derrida uses the word "corral" to evoke the image of animal captivity, and a couple of pages later he makes this image more vivid by speaking of this "catch-all concept" as a "confinement" and "encampment" (2008, 34). For Derrida, the definite article ("*the* animal") serves as a "strict enclosure" within which we corral different animals (2008, 34). But what kind of "enclosure"? Derrida gives a list of examples: "a virgin forest, a zoo, a hunting or fishing ground, a paddock or an abattoir, a space of domestication" (2008, 34). Derrida's list is surely dominated by instances of animal captivity that are at the service of human exploitation (zoo, hunting ground, abattoir). The slippage from the metaphoric enclosure (the definite article) to the literal enclosures that typify our treatment of animals is a reminder that the violence of reductive naming cannot be detached from the violence of our treatment of animals, and specifically the violence of animal confinement. The metaphoric confinement (of the concept) and the literal captivity inform and legitimatize one another.

In light of the slippage from the metaphoric to the literal, as well as Derrida's examples of enclosure, it seems that speaking of animals in a single voice always implies a certain kind of violence. Nonetheless, it would be a gross omission to overlook the very first example of enclosure offered by Derrida—a virgin forest. It is telling that Derrida juxtaposes a virgin forest with various artificial confinements. A virgin forest, insofar as it confines, also functions as a natural habitat. An abattoir, on the other hand, is an artificial confinement operated for the production of death. The contrast between the two perhaps suggests that we cannot even be sure of the nature of confinement. We cannot decide once and for all that a "confinement" is a site of violent exploitation. Indeed, the plot thickens when we consider not only the way we speak of, but also the way we *speak for*, animals. While the conceptual confinement of "the animal" is complicit in animal exploitation, it is the very same conceptual confinement that allows us to speak for them. The same "catch-all concept" of animality is also what animal advocates employ when they speak for animals.

The notion of "confinement" is further complicated in light of Derrida's deconstructive analysis of "hospitality." In various texts Derrida has articulated the ambivalent and even paradoxical traits contained within the concept of "hospitality." First, the French word "*hôte*" is ambivalent—it can mean both "guest" and "host" in English. (In his essay "Hospitality," Derrida sometimes uses English to specify which sense he means by "*hôte*.") Second, Derrida points out that the Latin root of the word "carries its own contradiction . . . [it is] parasitized by its opposite, 'hostility,' the undesirable guest" (2000, 3). Given the reversibility between guest and host, hospitality and hostility, it is not surprising that Derrida draws the further connection between hospitality and *hostage*: "The one inviting becomes almost the hostage of the one invited, of the guest [*hôte*], the hostage of the one he receives, the one who keeps him at home" (2000, 9). Some of us may have experienced the awkward moment when our guests overstayed their welcome. We are "held hostage" by our own guests, in our very own home. It is precisely because we are the hosts that we are the ones who are "held hostage"—after all, where else can we go?

The relationship between hospitality and hostage further complicates the notion of "confinement." Just as a host may be held hostage, a virgin forest may turn into a slaughterhouse. This is especially true when we consider the alarming rate of deforestation that leads to the loss of natural habitats for many species. As we have seen, Derrida includes zoos in his list of examples of confinement. Is the zoo a site of violent exploitation or is it a place where animals live long and prosperous lives? And what about our own homes? What happens when we welcome our animal companions to our home? Are we holding our cats and dogs hostage? Or is it the other way around? Consider, for example, animal caretakers who find it difficult to leave their animal companion at home alone. They spend more time at home and go on fewer trips once they adopt the animal. They are, so to speak, "held hostage" by the animal they welcomed home.

Finally, the ambivalence of "hospitality" challenges us to reconsider our responsibility to strangers. As we have seen at the beginning of this chapter, animal protection campaigns typically receive more attention from the public if the animals in question are "cute" or intelligent, and especially if they are perceived as harmless to humans. For example, more people care about elephants (ivory trades) and dolphins (captivity in amusement parks), even though the scale of their exploitation is much less than that of factory-farmed animals. So how can we extend our hospitality and ethical responsibility to animals that are "strangers" to us, especially when these stranger animals are "not cute," or even predatory?

If we take Derrida's deconstructive analysis of "hospitality" seriously, then it becomes clear that we cannot draw any easy distinctions between family and stranger, kin and enemy. As long as "hospitality" carries its own contradiction, there are strangers within one's own kin, and kinship is possible even with the strangest strangers. Consider, for example, the incredible story of wildlife conservation in Namibia. In the 1980s and 1990s, conservation organizations began paying locals to take care of wildlife. The locals who best knew the wildlife and the flora turned out to be poachers. So these poachers began taking care of the animals they used to hunt. As a result of the conservation efforts by these former poachers, many animal species at the brink of extinction have increased in numbers in recent years.[21] The reversibility between poachers and caretakers (strangers and family) means that we can never be certain of the *limits* of our ethical responsibility and hospitality.

## SILENCING THE *ANIMOT*?

In response to the reductive violence of the name "animal," Derrida coins the word "*animot*"—a neologism that seeks to reestablish plurality in "the animal" (the singular). Because "*animot*" is a homonym of "*animaux*" (animal*s* in French), "*animot*" is heard as a plural or, more precisely, the plural is heard in the singular. In addition to the emphasis on plurality, "*animot*" also contains the word "*mot*," the French word for "word." By inserting "*mot*" into the animal, Derrida mocks the philosophical cliché that posits the human as the speaking animal, the linguistic divide that has hitherto deprived animals of words. But "*animot*" does more than challenge the typical view that animals lack language. For it also reminds us that, by depriving animals of language, we are at the same time depriving language of animation. As Laurel Peacock puts it,

> The term *animot* can be thought of as marking an animal invasion of the word (*mot*), in which animal otherness animates language. An *animot* is an animalistic kind of word, and a linguistic kind of animal, attributing animation, even agency, to language. Thus on the level of the word, and even on the level of the letter, animal otherness can be invited into language. (2009, 89–90)

But perhaps it is not an animal *invasion* of language, as "invasion" implies a transgression of territory. Rather, "*animot*" seems to *uncover* the animality that had once animated language.[22] In fact, even Peacock seems to agree that animality is always already part of language, as she immediately reminds us that the first letter of the English alphabet is an "iconic transcription of an animal into language"—the alpha of the alphabet comes precisely from the representation of an ox's horn and its triangular face, upside down.[23] Our language has always been indebted to animals.

It is significant that the word "*animot*" is *heard* as a plural. Insofar as the neologism depends on the vibration of our vocal chords to take effect, the diversity of *animot* is manifested only when it is spoken/heard. (And let's not forget that *The Animal* was first given as a lecture.) Put differently, the inscription must be substituted by the voice for the plurality to come through. However, as long as the effect takes place only in *speech*, only by saying it out loud, is Derrida inadvertently privileging speaking? To complicate this substitution further, in his lecture Derrida makes the following demand:

> In order not to damage French ears too sensitive to spelling and grammar I won't repeat the word *animot* too often. I'll do it several times but each time that, henceforth, I say "the animal" [*l'animal*] or "the animals" [*les animaux*] I'll be asking you to silently substitute *animot* for what you hear. (2008, 47)

Given that the plurality of *animot* hinges upon it being a homonym of *animaux*, it is particularly strange that Derrida asks his audience to "silently substitute" his neologism for "animal/animals." Apart from the technical difficulty of delivering a sound effect in silence, Derrida's reluctance to repeat the word *animot* is troubling. For how can a new lexicon deliver its critical force if it could only be "silently substituted"? Furthermore, given that in the history of Western thought the animal is repeatedly posited as lacking speech, as *silent*, what should we make of Derrida's request to make a silent substitution? Paradoxically, while Derrida's neologism hinges on its vocalization to take effect, its delivery must be made in silence.

Issues of substitution continue to multiply when it comes to the translation of *The Animal*, in particular when we consider the fact that the play on homonyms becomes effective only in French. There seems to be no English substitute for *animot* that would deliver its playfulness as well as its nuances. Even Derrida himself readily admits that "*animot*" is "more untranslatable than ever" (2009, 112n22). In the English edition of the lecture, "*animot*" is left untouched, untranslated—and uncontaminated. While keeping a technical and crucial word in its original language is a typical practice in translation, does it also betray a certain reluctance to interrogate the English lexicon? By leaving Derrida's neologism untranslated, we acknowledge that "*animot*" is a singularity irreplaceable by any existing English word(s). But it can also mean a missed opportunity to expand the existing vocabulary. Is it merely a deferment of an interpretive decision? (But doesn't a deferment also involve a decision?) Or is there perhaps an unwillingness to adulterate Derrida's neologism?

In *This Is Not Sufficient*, Leonard Lawlor addresses the issue of purity in translation. Responding to Derrida's confession of his taste for purity in language, Lawlor makes the following observation:

> The idioms of a language are what make the language singular. An idiom is so pure that we seem unable to translate it out of that language. Derrida, we have seen, always connects the French idiom "il faut," "it is necessary," to "une

> faute," "a fault," . . . but we cannot make this connection between necessity and a fault in English. (2007, 118)

Just as the necessity-fault connection cannot be replicated in English, the *animot-animaux* play is exclusive to the French language. With Derrida's extensive use of word play, translation often seems impossible. But this impossibility also reflects a demand that we make of translation. It is a demand for preserving purity; that is, the purity of the original. Because words carry their own histories and cultures, translation often threatens to contaminate the original.

It is against the backdrop of this demand for purity that Lawlor argues that in Derrida's view "what is most pure in a language . . . is the very possibility of impurity" (2007, 119). Lawlor points out that the ambiguity of idioms renders translation necessary even within the same language. In his example, the French idiom "*il y va d'un certain pas*" could either be about a movement to a place at a certain pace or about the issue of negation; and such undecidability shows that "already in the French, in the one French language, there is already translation" (2007, 119). But if impurity is at the heart of the pure and the original, what should we make of the (alleged) unsubstitutability of *animot*? What is it that we are trying to preserve by keeping the word "*animot*" in an English translation? If Derrida's neologism is supposed to showcase the possibility of a linguistic disruption, then it is rather ironic that we fossilize his word in its original French, foreclosing the possibility of rejuvenating and complicating the English lexicon. Insofar as Derrida's *animot* is supposed to be fruitful and multiply (within the concept of animality as well as the French vocabulary), we have inadvertently neutered Derrida's *animot*. In the end, grafting the word *animot* in the English translation may have undermined the significance of Derrida's neologism.

## TRANSLATION AND THE PURITY OF ANIMAL VOICE

The question of translation brings us back to the question of speaking for animals. As I argued in chapter 1, the tendency to valorize the voice of animals (by insisting that animals should speak for themselves) betrays the fantasy that there is a pure animal voice that speaks through us, uncontaminated by human interests. However, if impurity is at the heart of the pure and the original, then what is this impurity in animal voice? As we have seen, there are multiple—and conflicting—animal voices. The heterogeneity of animal voices makes translation both impossible and inevitable. Translation is impossible insofar as this heterogeneity of animal voices is irreducible to a singular interpretation. But translation is also inevitable insofar as this multiplicity necessitates our selective listening of particular voices. After all, the fact that we cannot capture the multiplicity of animal voices does not mean that we hear *nothing*; rather, it means that what we hear is already filtered through our interests and biases. In other words, the impurity in the animal voice means that we must decide the undecidable.

In *The Beast and the* Sovereign, volume 1, Derrida calls into question the purported uniqueness of the human. He maintains that it is less a matter of whether the animal actually lacks the capacity that we refuse them, but more a matter of whether "man has the right, for his own part, to attribute in all rigor to man, to attribute to himself, then, what he refuses to the animal" (2009, 130). In keeping with the Derridian spirit of interrogating our supposed uniqueness, we should also ask, can *we* speak for ourselves? Do *we* have a voice that we can properly call our own? In his article "Following the Rats: Becoming-Animal in Deleuze and Guattari," Lawlor once again concludes with a reflection on purity and language. There Lawlor examines the experience of auto-affection, specifically whether it is "truly 'auto,' uncontaminated by any other" (2008, 182). He remarks his experience of internal monologue as follows:

> It is an irreducible or essential necessity that the silent words I form contain repeatable traits. This irreducible necessity means that when I speak to myself, I speak with the sounds of others. In other words, it means that I find in myself other voices . . . there is a memory of multiplicity, of the many voices that are in me. Thus the problem with the belief that interior monologue is my own is that others' voices contaminate the hearing of myself speaking. (2008, 183)

If other voices infect even one's internal monologue, can one still claim a voice to be one's own? As speaking beings, we are always already implicated in a linguistic structure not of our own making. To state the obvious, in order to speak we must employ existing vocabularies and follow the regulation of grammatical rules and conventions. As we have seen in chapter 1, our judicial system requires the mediation of a human delegate when it comes to enforcing animal rights. Insofar as the human delegate speaks in legal lingo and not in an animal voice, the animal does not speak for itself. However, the practice of speaking for animals is symptomatic of the representational structure on which our judicial system rests. After all, do we get to speak in our own voice in court? Aren't most plaintiffs/defendants *represented* by their lawyers? And so long as what we want to say is always channelled and regulated through the voices of others (especially those who are fluent in legal language), what does it still mean to speak in our *own* voice?

Issues of translation not only complicate the rhetoric of speaking for animals, they also challenge our certainty of having a voice of one's own. But perhaps the dubious opposition between purity and contamination is just a different expression of the dubious opposition between identity and difference. After all, if there is no identity so pure that it escapes difference, then there is also no voice that is purely one's own, uncontaminated by the voices of others. In the end, just as difference is at the heart of Wittgenstein's "family resemblance," the "hetero" is also at the heart of the "auto."

## NOTES

1. All references to state laws come from Stephan Otto's sixth edition of the "Animal Protection Laws" compendium (2011).
2. Indeed, even the meaning of "pests" in the legal realm is tenuous and it does not always conform to our common notion of "pest." For instance, the peacock is considered a kind of pest. In 2011, an early version of the "Peacock Bill" (SB 1533) sought to stipulate that peacocks do not count as pests. But the final version of the bill (passed in Hawaii) dropped this stipulation and merely states that the killing of the pest should be "handled in accordance with standard and acceptable pest control practices and all applicable laws and regulations."
3. Francione (2011).
4. An organization called "Research Saves" argues that animal researches save lives—both human and animal. See http://www.researchsaves.org/.
5. This is the case for those who adopt the abolitionist approach, the approach that rejects all use of animals, but it is not necessarily the case for those who adopt the welfarist position.
6. Medina's essay, "Identity Trouble: Disidentification and the Problem of Difference" (2003), offers a succinct account of the problem of group identity.
7. See Wise (2002, 207–30); Fellenz (2007, 46). For a general account of how philosophers deal with the linguistic divide, see Steiner (2005, 18ff.).
8. See Gruen (2011, 6ff.)
9. See Shapiro (2006).
10. Wittgenstein (1967, 518).
11. Wittgenstein (2009, 250).
12. Wittgenstein (2009, 363).
13. Wittgenstein (2009, 327).
14. Wittgenstein (2009, 314).
15. In his obituary of Wittgenstein, Bertrand Russell (1951) recounts the story that Wittgenstein refuses to "believe" the statement that "There is no hippopotamus in this room at present." It is interesting that the animal in question becomes a hippopotamus in this version of the story. In one of Russell's letters in which he recounts the same story, the animal is a rhinoceros (Monk 1990, 39).
16. R. G. Frey (1980) and Michael Leahy (1991) both use Wittgenstein to advance their anti-liberationist stance, while David DeGrazia (1994), Nigel Pleasants (2006), and Wendy Lee-Lampshire (1995) all find Wittgenstein helpful for reconceiving our relationship with animals.
17. Wittgensteinian scholars such as Cora Diamond and Stephen Mulhall have contributed important work on animal ethics. See Diamond (1978; 2001; 2003) and Mulhall (2009). (While Diamond does not explicitly invoke Wittgenstein in her 1978 essay "Eating Meat and Eating People," she does so in "Injustice and Animals" [2001]. Specifically, she employs Wittgenstein's notion of "difference in grammar" to articulate different senses of justice [119–20].)
18. For a relevant discussion of the problem of difference and identity politics, see Wendy Lee-Lampshire (1995). In it, Lee-Lampshire articulates a brand of ecofeminism in the Wittgensteinian vein. Specifically, she argues that Wittgenstein's "naturalistic" conception of subjecthood is helpful for those who seek to articulate a feminist standpoint, but are wary of the essentialist baggage that often comes with standpoint theory. In her view, a Wittgensteinian-inspired ecofeminist standpoint "takes as its point of departure the complexities and dissonances which characterise the bio-psycho-social positions occupied by thing" (1995, 99). For her, such a "standpoint" is not a unified position; rather, it recognizes conflicts and differences as its inherent and ineliminable qualities.

19. For example, we may think that there is something amiss if a father tells his children that he takes care of them because they have the *rights* to be fed, clothed, and sheltered, rather than because he cares about them.

20. Worse still, family violence is sometimes not even *recognized* as violence (or punishable violence). For instance, up until the 1970s, spousal rape was not considered a crime in most states in the United States. Even though now all fifty states have criminalized spousal rape, spousal rape is still not treated as equal to non-spousal rape. Specifically, some states recognize spousal rape as a crime *only if* the aggressor used or threatened to use force (Tennessee, for example, recognizes spousal rape as rape only if the aggressor was armed, carried a credible decoy, or if the victim is seriously injured). Such additional clauses make it more difficult for spousal rape to *count* as rape.

21. Kasaona (2010).

22. Abram offers a helpful account of how human language is indebted to the larger non-human world in his book, *The Spell of the Sensuous* (1996). See especially chaps. 3 and 4.

23. Peacock (2009). Indeed, the letter "A" is not the only letter that takes after the animal. The letter "Q" also supposedly comes from the representation of a monkey (imagine the tail) (Abram 1996, 101). In *Animal Lessons*, Kelly Oliver makes a similar point regarding the relationship between animals and language by noting that "the first ink used was animal blood . . . some of the first pens were bird quills, and . . . animal figures were some of the first written 'language'" (2009, 118). Notably, the oldest form of Chinese characters, the "Bone Oracles," are inscribed on turtle shells.

# *Chapter 5*

# Racializing Cruelty

## *Dehumanization in the Name of Animal Advocacy*

In previous chapters, I argued that speaking for the animal is a responsibility that we cannot disavow. But if we cannot avoid representing animals, then we need to find a way to *better* represent them. In this chapter, I argue that we cannot speak well for the animal unless we take intersectionality seriously. Just as ecofeminists have interrogated the ways in which speciesism and sexism perpetuate one another, we need to examine the ways in which speciesism and racism inform one another.

In 2010, English musician Morrissey stirred controversy after attacking China's animal welfare record during an interview with *The Guardian*. In the interview, the musician, whose second solo album is titled *Meat Is Murder*, expresses his disgust over the Chinese people's treatment of animals as follows: "Did you see the thing on the news about their treatment of animals and animal welfare? Absolutely horrific. You can't help but feel that the Chinese are a subspecies" (Armitage 2010). Anti-racism groups such as Love Music Hate Racism immediately censured Morrissey's remark. The spokesperson of the group condemned Morrissey's remark as "crude racism" and made clear that the group could no longer accept Morrissey's donation unless he rescinded his comment.

Although the controversy revolves around the overt racism of Morrissey's remark, the language of "subspecies" is interesting in its own right. Who or what counts as a member of this "subspecies"? What is this "proper species" under which the "subspecies" resides? Indeed, what is the ethical-political force of indicting the Chinese as "subspecies"? Of course, what Morrissey meant to say is that the Chinese are so inhumane in their treatment of animals that they are beneath what we consider proper to the human. In other words, by mistreating animals, the Chinese have become *less than human*. Morrissey's remark—while intended to speak for the animal—curiously invokes and reasserts the human-animal hierarchy. But what is this double movement of speaking for the animal while simultaneously resorting to the human-animal hierarchy? How effectively can we speak for the animal if we continue to rely on the language of "subspecies" or "subhuman"—a language that presupposes human exceptionalism?

The Morrissey controversy helpfully highlights two theses that I want to examine in this chapter: (1) animal advocacy can be employed to reinscribe white privilege, or even to further racist and xenophobic agendas, and (2) human superiority reasserts itself in animal advocacy. Of particular interest is the ways these two theses overlap. Although these two theses have been articulated and problematized independently in the literature, their *connection* remains to be examined. There is a pressing need to examine how animal advocacy has become a site where racism and speciesism feed on each other, and it is the connection between the two that I aim to establish in this chapter. It is by taking the link between racism and speciesism seriously—rather than treating it as merely analogical—that we can begin to interrogate some of the problematic assumptions in animal advocacy and begin speaking better for animals.

I will begin by examining the connection between racism and speciesism in early animal liberation discourse. Despite the fact that racism has served as a compelling analogue to illustrate the wrongness of speciesism and human exceptionalism, there is a danger with this analogy—the overemphasis on how the two are analogous gives one the distorted impression that they are merely parallel to each other. However, it is important that we look at how racism and speciesism *intersect*, rather than how they merely resemble each other. As recent feminist posthumanist discourse forcefully demonstrates, animal advocacy can reinscribe white privilege and be complicit with the oppression of racial minorities. Following their lead, I too will interrogate how animal advocacy can become complicit with racist agendas; however, I will argue that this complicity is informed by the human exceptionalism that is still at work in animal rights discourse. In other words, I argue that animal advocacy is susceptible to racist and xenophobic agendas because animal advocacy continues to be speciesist. I will shed light on how the intersection of racism and speciesism renders animal advocacy especially vulnerable to racist agendas. As I will show, one important contributing factor is the problematic link between cruelty and animality: as long as cruelty is associated with animality, those who commit acts of cruelty against animals will continue to be animalized. I end this chapter by examining how racism, cruelty, and animalization enter into a triangulated relationship in American anti-immigration politics.

## RACISM AND SPECIESISM: THE ANALOGY DEBATE

There has been a long history of racist discourse in which the racial other is compared to, or associated with, an animal. Analogizing the animal and the racial other is meant to "dehumanize" the racial other, so naturally one strategy to reject this animal trope is to reject the analogy between the animal and the racial other. But this language of "dehumanization" already presupposes a human/animal hierarchy: it is bad to be compared to an animal because we are prejudiced against the animal; it is worse to be treated like an animal because we treat animals cruelly. It is easy to see why animal advocates would find the rhetoric of dehumanization problematic. So can we curb racism without throwing the animal under the bus? Is there a strategy with which the racial other and the animal other may stand together, rather than in opposition? Enter "speciesism," an *inversion* of the original animal-race analogy, popularized by Peter Singer in his book *Animal Liberation*.

In *Animal Liberation*, a foundational text for the animal liberation movement, Singer identifies "speciesism" as the crux of the injustice that defines the human exploitation of animals. He makes his case against speciesism *by way of* arguing against racism and sexism: to see why denying equal consideration to animals is wrong, "we need to see, *first*, exactly why racism and sexism are wrong" (2009, 3; emphasis mine). Singer proceeds to explain that racism and sexism violate the principle of equality insofar as genetically based differences should not affect one's claim to equality. Speciesism, *like racism and sexism*, violates the principle of equality by allowing the interests of our own species to override the greater interests of members of other species (2009, 4–6). As such, if we don't want to be a racist or a sexist, we also wouldn't want to be a speciesist. But as we will soon see, even though Singer anchors his argument against speciesism in the wrongness of racism and sexism, animal advocacy can still be appropriated by, and complicit with, racist discourse.

Before we turn to the ways in which animal advocacy has been at the service of racism, it is important to note that even though Singer's analogy has proliferated in animal rights discourse, it has been contested as well. For example, Carl Cohen calls the comparison between speciesism and racism "insidious" (Cohen and Regan 2001, 62). In his view, racism is unjustified because "humans *really* are equal," whereas speciesism is a "correct moral perspective" insofar as there are "morally relevant differences" between humans and other species of animals (2001, 62; emphasis mine). Leslie Francis and Richard Norman (1978) are concerned with the political implications of comparing animal advocacy to the civil rights and women's liberation movements: "the equation has the effect of trivialising those *real* liberation movements, putting them on a level with what cannot but appear as a bizarre exaggeration" (1978, 527; emphasis mine). According to these critics, only human liberation movements are "real," presumably because only human equality is "real." For many, animal suffering and human suffering appear so incomparable that the comparison of the two is deemed offensive. In 2009, Germany's High Court ruled against an ad campaign run by People for the Ethical Treatment of Animals (PETA) that involved a comparison of animals in the slaughterhouse and Jews in concentration camps. The rationale for banning the ad campaign was that it "would have made the fate of the victims of the Holocaust appear banal and trivial" ("High Court" 2009). This comment reveals how we value an animal's life: the comparison makes the fate of the Holocaust victims appear "banal and trivial" only because the death of an animal is considered "banal and trivial" in the first place.

While it is certainly important to consider the implications of the speciesism-racism analogy, the analogy between speciesism and racism is only one way to articulate their relationship, and the relevance of racism in animal ethics is more than just to demonstrate the wrongness of speciesism. In the following sections, I will consider the link between racism and animal advocacy that goes beyond the analogy debate.

## REINSCRIBING WHITE PRIVILEGE THROUGH ANIMAL ADVOCACY

Though feminists have long considered the animal question through the lens of gender inequality and oppression, the intersection between race/culture and animal ethics is a relatively recent effort (Deckha 2012).[1] Fortunately, given the rich connection between animality and race, there is a burgeoning conversation examining how species and race intersect in animal studies.[2] This "border crossing," as A. Breeze Harper calls it, considers not only how speciesism and racism are dependent on each other, but also how speciesism is invoked to protect white identity in popular discourse (2011, 75).

Outside of the rights discourse, feminists have offered rich analyses of the relationship between racism and animal advocacy. In her essay on the Makah whale hunt, Greta Gaard (2001) examines the tension between ecofeminists' commitments to whales and their respect for the Makah tradition. She presents the dilemma by asking: "how do we voice our dissent about the oppressive features of traditional cultural practices in a way that does not reinscribe colonialism, enhancing divisions within the tribe for our own commercial or political purposes?" (2001, 18). That is, what should ecofeminists do when their ecological commitments appear to be at odds with their anti-colonial sensibilities? Gaard articulates the connection between racism and animal advocacy not by analogizing racism and speciesism, but by showing that animal advocacy may in fact be complicit in racism and colonialism. In a similar vein, Cathryn Bailey examines the link between animal advocacy and white identity. In her 2007 essay "Africa Begins at the Pyrenees," Bailey points out that moral indignation over bullfighting has become a trope that demarcates the "civilized" and the "uncivilized." The rhetoric against bullfighting is imbued with racist and colonialist language:

> It has often been argued that bullfighting was further proof that Spain was actually more Eastern than Western, or more African than European. According to Adrian Shubert, "The true horror of the bullfight was that it turned Spaniards from Europeans into Africans. . . . The bullring brought Spaniards down to the level of the Moors, and the bullfight was nothing more than an 'African ferocity.'" (2007a, 31)

Similarly, in another 2007 essay Bailey examines the connection between diets and the production of racial identity (2007b). She points out, for example, that fetishism for ethnic food has become a way for middle-upper-class whites to redefine their whiteness. By adopting exotic ethnic food or alternative diets, middle-upper-class whites can rearticulate their identity as a "special" whiteness; specifically, they distinguish themselves from other "reviled suburban" whites, whose unsophisticated palates render them quite content with supermarket frozen meals and chain restaurants. Her analysis reveals that one's identity (including racial identity) is intimately tied to one's diet.

Several recent essays examine the role of race in Michael Vick's dogfighting scandal. In "The Dangerous Individual('s) Dog," Erin Tarver argues that the use of non-human trope in racist discourse goes beyond analogizing the animal and the racial other. The racial other is deemed "dangerous" by being *associated with* a dangerous animal (in this case, pit bulls). This "danger" is twofold: the supposed criminality of the dangerous individual, as well as the contagiousness of this criminality. The danger of the racial other and the animal feed on each other: the racial other is dangerous because of his dangerous animal, and it is only because he is a dangerous individual that he would keep such dangerous animals (2013, 10).

In "The Wonderful, Horrible Life of Michael Vick," Claire Jean Kim uses the dogfighting case to highlight the importance of intersectional analysis (2014). She argues that discourse surrounding the scandal fails to consider how the race question and the animal question intertwine. In Kim's analysis, Vick's critics tend to reduce the case to an issue of animal welfare by assuming a race-neutral, colour-blind position, while Vick's defendants tend to silence the animal question by highlighting the racial double standard. It is as if one could either fight speciesism or racism, but not both.[3]

Gaard, Bailey, Tarver, and Kim showcase a fruitful way to reconsider the connections amongst race, racial identity, vegetarianism, and animal advocacy. Specifically, they are attuned to the ways animal rights rhetoric may reinscribe white privileges. My analysis of racism and animal advocacy in this chapter is similar to their approach insofar as I, too, am concerned with the ways animal rights rhetoric may be used in the service of racist and imperialist discourse. However, my analysis also departs from theirs. I am especially concerned with how the human-animal hierarchy reasserts itself at the very moment when animal advocacy and racist discourse intersect. To paraphrase Gaard, I ask, how do we voice our dissent about the oppressive features of traditional cultural practices in a way that does not reinscribe *both* colonialism *and* human exceptionalism?

Before I move on, I should note that my motivation to interrogate the complicity of animal advocacy in racist and imperialist discourse is neither to undermine the importance of animal advocacy, nor to imply that we should care *more* about racism than speciesism. If my analysis is correct in that racism and speciesism feed on each other in animal advocacy, then we cannot address one without the other. As Kim argues persuasively, fighting racism and speciesism are not mutually exclusive (2014; 2015). In fact, the failure to acknowledge their interconnectedness results in what she calls "single-optic vision," which is "a way of seeing that foregrounds a particular form of injustice while backgrounding others" (2015, chap. 1). Animal advocates activate the optic of cruelty when they downplay the role race plays in their advocacy, especially the ways in which racial hierarchy is being reproduced when they speak for the animal. Minorities activate the optic of racism when they downplay the significance of animal suffering by reasserting human dominance over animals. This single-optic vision leads to "a posture of mutual disavowal—an explicit dismissal of and denial of connection with the other form of injustice being raised" (2015, chap. 1).

Instead of operating on this single-optic vision, Kim argues that we need a "multi-optic vision" to develop an "ethics of avowal" that recognizes the connection of multiple forms of injustice. Racism and speciesism are not simply analogous to each other, as if they are two forms of struggle paralleling each other but never intersecting; nor are they antithetical to each other, as if we can only fight one but not the other. Rather, the oppression of racial/cultural minorities and the oppression of animals are linked because they feed on each other. The fight against racism and the fight against speciesism is not an either/or, but a both/and. I will return to the question of cultural diversity and animal advocacy at the very end of this chapter.

## ANIMALIZATION AS DEHUMANIZATION

It is no news that racist discourse often invokes the animal. Dehumanization is often achieved by *animalizing* the other. Calling a person by an animal name or comparing that person with an animal is injurious because it is intended to strip the person of the dignity that is proper to the human. More importantly, animalizing a person makes it easier to *treat* the person like an animal.[4] For example, when anti-Semites call Jews "vermin," they reinforce the idea that all Jews should be exterminated. The strategy of animalizing the other presupposes human superiority: animalization is viewed as "dehumanizing" because being human means *not being an animal*. It is bad to be an animal, good to be a human. As such, dehumanization derives its power from the human-animal hierarchy that is already in place.

In their essay "Le Pratique Sauvage: Race, Place, and the Human-Animal Divide," Elder, Wolch, and Emel (1998) identify three major ways that the animal body has been used as a tool of dehumanization. The first two are familiar: people are dehumanized when they are either treated like animals or when they are compared to animals (e.g., the comparison of blacks and apes[5]). In addition to these comparisons of likeness, racist rhetoric often highlights the danger and foreignness of the racial other by associating them with *diseased* animals. The 2009 swine flu epidemic offers a telling example: during the early phase of the outbreak, Israeli Health Minister Yakov Litzman complained that the reference to pigs is "offensive" to both Islam and Judaism, and that "swine flu" should be renamed "Mexican flu" ("Israel Renames Unkosher Swine Flu" 2009). The easy slippage between "swine" and "Mexican" is far from innocuous. As physician Stanley Aronson points out, assigning geographic names to communicable diseases segregates the world into the "contaminated and the uncontaminated" (2010). And such line drawing is informed by a "racism which . . . is prompted by the inchoate fear that the third world is intent on sending both its uneducated young and its threatening pathogens to seek shelters on our pristine shores" (2010). Indeed, the swine flu outbreak fuelled discussions on border control, and anti-immigrant groups predictably exploited the epidemic to push their agenda. Michelle Malkin blogged on her website that "the spread of contagious diseases from around the world into the U.S. [is] a result of uncontrolled immigration" (2009). Talk show host Michael Savage warned his listeners that the swine flu could be a terrorist conspiracy, as virus-carrying Mexicans were marching across the U.S. border as humanoid biological weapons (Alexander 2009). The slippage between "swine" and "Mexican" feeds the racist imagination whereby the foreigners are crossing borders to take over our land, much like the swine is crossing the species borders to take over our body.[6]

The swine/Mexican flu example shows that animalization has become a typical tactic by which immigrants are racialized; there is an ineliminable link between animalization and racialization. This brings us back to Elder, Wolch, and Emel's analysis of dehumanization. In addition to comparing and treating someone like an animal, Elder, Wolch, and Emel argue that people are dehumanized when their *treatment of animals* is deemed savage and uncivilized (1998, 82ff.). Focusing on the ways subaltern "animal practices"[7] become the site of racial conflicts, Elder, Wolch, and Emel show that animal cruelty has been used by dominant groups to racialize and marginalize immigrant groups. The dominant group calls attention to the otherness of immigrants not by presenting them as beasts, but by presenting them as "people-acting-*beastly* toward animals" (Elder, Wolch, and Emel 1998, 82; emphasis mine). The racial other is portrayed as an animal not because he looks like or is treated like an animal, but because he acts like an animal *by being cruel to animals*.

As we have seen, recent feminist discourse has cautioned us against the danger of reinscribing white privilege via animal advocacy. Animal cruelty has long been used to mark the "uncivilized" other; we can trace this genealogy to the time of colonial dominion when foreign missionaries and colonial governments took up the "white man's burden" of civilizing and moralizing local "savages." For example, Hong Kong remains the only Chinese region[8] where the slaughtering of dogs and cats is prohibited—thanks to the British colonial government instituting the "Dogs and Cats Ordinance" in 1950. The fact that it is the dogs and cats (rather than the cows, pigs, or chickens) that receive protection shows that the ordinance has more to do with the British sensibilities for certain domesticated pets than animal welfare. As Maneesha Deckha points out, early anticruelty laws in Britain typically targeted practices associated with the lower classes (e.g., blood sports), rather than the upper classes (e.g., fox hunting) (2013, 521). She argues that this double standard in domestic anticruelty law is more than just an incidental hypocrisy: by targeting animal abuse associated with the lower classes, "the superior 'national' identity of England could be articulated vis-à-vis colonial societies without jeopardising Victorian animal-based practices" (2013, 521). In other words, civilizing the lower classes in the motherland goes hand in hand with moralizing the "savages" in the colonies.

While using animal cruelty to mark the racial other has a long colonial and imperialist history, the analysis offered by Elder, Wolch, and Emel is particularly instructive as it foregrounds the human exceptionalism at work in racist discourse. The racial other is deemed uncivilized because of their cruel treatment of animals; their cruel treatment of animals, in turn, signals their *animality*, making them less than human. In other words, the very censure of their animal cruelty continues to invoke the human-animal hierarchy. While in their paper Elder, Wolch, and Emel focus primarily on immigrants and minorities born in the United States, their argument is also instructive for the larger international context. Following and building on the insights of Elder, Wolch, and Emel, I argue that *The Cove*, a 2009 documentary, unwittingly reinscribes the human-animal binary despite its effort to advocate for the animal.

## CASE STUDY: *THE COVE*

In "Fishing for Animal Rights in *The Cove*," Carrie Freeman argues that despite its animal rescue mission, *The Cove* inadvertently reinforces the human-animal hierarchy that privileges the human. She maintains that the filmmakers repeatedly highlight the superior cognitive capacity of dolphins in order to elicit sympathy and respect from the audience. At different junctures the filmmakers attribute to dolphins an array of qualities that we typically consider human: intelligence, self-awareness, language, and, perhaps the most surprising of all, a suicidal proclivity when life is no longer bearable. According to Freeman, the documentary "reinforces a humanist bias" (specifically a "cognitive speciesism") when they single out dolphins as worthy to be rescued by virtue of their likeness to humans (2012, 114).

Freeman's critique follows the long line of criticism of the strategy of identification in animal ethics literature—why should animals be *like us* in order to deserve moral consideration? Why should we care more about this particular species of animal just because they are intelligent and self-aware? Human exceptionalism is reinscribed in animal advocacy as long as we believe that animals are worth saving only when they possess certain human-like qualities. Though Freeman's application of this critique to *The Cove* is helpful, the critique itself is not new.

In what follows, I will also argue that *The Cove* resurrects the human-animal binary and its concomitant human exceptionalism. However, my argument departs from the typical critique that Freeman reproduced. Whereas the typical critique focuses on how we elevate certain animals because of their human-like qualities, my argument focuses on how we denigrate certain humans by attributing animal-like qualities to them. Specifically, my analysis centres on how the supposed human villains in the documentary are portrayed as less than human. Just as the dolphins are *humanized* by taking on qualities that we consider human, these villains are *animalized* by taking on qualities that we consider "beastly."

*The Cove* received considerable attention for capturing footage of dolphin slaughtering in Taiji, a remote Japanese fishing village. Even though it was the gruesome footage that made the film famous, a major part of the film is devoted to the making of the footage. In one interview, director Louie Psihoyos notes that the "making-of" portion of the documentary is reminiscent of a thriller:[9] the covert mission of capturing the footage was planned in a closed hotel room, performed in the dark, facilitated by various decoys, and chronicled by a military infrared camera. This environmental *Mission Impossible* is peppered with high-tech props such as camouflage, hydrophones, thermal-grade cameras, and even "rock cams" (fake rocks that hold the secret cameras). But one element that makes *The Cove* a classic thriller—an element that Psihoyos neglects to mention—is the hero-villain opposition. If the supposed "hero" in the film is the team of dolphin activists, the "villains" are those who support the dolphin fishing business. Unlike the high-tech activists, the Japanese fishermen are portrayed as barbaric and dangerous—especially given their low-tech and bloody way of killing the animals. In one particularly memorable scene, a Japanese fisherman kills dolphins in the cove by stabbing them repeatedly with a spike as the seawater turns from blue to a deep red.[10]

Recall that Elder, Wolch, and Emel argue that people are dehumanized and marginalized when their treatment of animals is being scrutinized as savage and uncivilized. Following their analysis, the brutal death that the Japanese fishermen inflict upon the dolphins is indicative of their otherness. The Japanese fishermen are portrayed in the documentary as cruel and violent—the harpooning-dolphins-to-death scene at the end makes them look particularly "savage." The main activist, Ric O'Barry, implicates the Japanese mafia as the supporter and beneficiary of the dolphin fishing business, thereby imputing the image of gangster violence to the Japanese fishermen as well.[11] As Elder, Wolch, and Emel point out, the dehumanization that is operative goes beyond treating the subaltern group (in this case the Japanese fishermen) as animals; it also goes beyond attributing likeness between the group and the animal. Rather, the dehumanization of these Japanese fishermen is achieved precisely by *pitting them against* the animal. It is their treatment of the animal (their practice) that dehumanizes them, alienating them from the dominant, superior group.

According to Elder, Wolch, and Emel, shifting the focus to *practice* is distinctive of the postcolonial and postmodern epoch. They argue that this shift results from "radically changing time-space relations that epitomise postmodernity" (1998, 81). The compression of space "brings visible difference 'home' instead of restricting it to a distant, exotic colonial space" (1998, 82). That is, whereas in the colonial past, the "us" and "them" were geographically segregated, in postmodern times the "us" and "them" are often dwelling in the same place due to the hyper-mobility of globalization. As such, the exotic practices of the foreigners are no longer fantastical stories that one reads from a missionary's journal, but realities that one must confront. In the case of *The Cove*, even though the slaughtering of the dolphins takes place in a distant Japanese village, the compression of space is still operative. *The Cove* brings to light the "exotic" practice behind the seemingly innocuous American pastime—dolphin entertainment at water parks such as SeaWorld. The space between the Japanese cove and the American audience is compressed when the audience realizes that the horrific slaughter in a faraway cove is happening in large part because of the demand for dolphin entertainment at home (as dolphins that are unfit for sale are slaughtered). While the slaughter is physically taking place in Japan, the heartland is implicated.

While the changing time-space relation offers one explanation for this new focus on animal practice, the posthumanist challenge to the centrality of the human should not be overlooked.[12] The project of decentring the human subject that characterizes the posthumanist movement disrupts the place of the human as well as the way we interpret the world. However, if the human is defined against the animal, then the displacement of the human would also muddle the human-animal hierarchy. Our heightened sensitivity to the *treatment* of animals is reflective of the posthumanist scepticism towards anthropocentrism—the infrastructure that has hitherto grounded our relation with animals. In a sense, the animal turn in academia and the proliferation of public discourse on animal welfare are indicative of our posthumanist sensibilities. But as we will see, even the disavowal of anthropocentrism continues to rely on anthropocentrism. It is important to consider the following questions: Do we succeed in disrupting the human-animal hierarchy by challenging practices that we deem anthropocentric? Or are we in fact reinscribing the human-animal hierarchy in the name of animal advocacy? To begin answering these questions, we need to examine the connection between cruelty and animality.

## BEASTLINESS AS A GIVEN OF THE BEAST

What is the connection between cruelty and animality? It is not uncommon to hear heinous crimes being described as "beastly," "brutal," and, most tellingly, "*in*human." There seems to be a link between beastliness and cruelty, at least in common parlance. The etymological kinship between beastliness and beast, brutality and brute, humane and human is revealing: beastliness is set in opposition to humaneness just as the brute is set in opposition to the human. Beastliness is antithetical to humaneness and humanitarianism. "Beastly" is what we become when we fail to act humanely—that is, when we fail to act kindly and compassionately.

Setting aside these etymological connections, the link between cruelty and animality can also be found in the history of philosophy. Seneca, for example, considers cruelty a mark of animality:

> Cruelty is far from being a human vice, and is unworthy of man's gentle mind: it is mere bestial madness to take pleasure in blood and wounds, to cast off humanity and transform oneself into a wild beast of the forest. (Seneca, book 1, xxv)

Human cruelty, in Seneca's account, is a reminder that we have sunk to the level of the "wild beast." Cruelty is unworthy of our mind because we are supposed to transcend our animality.

The idea that cruelty makes one a beast is also in J. M. Coetzee's novel *The Lives of Animals.* In the story, the main character Elizabeth Costello calls attention to the metaphor of beastliness in narratives about Nazi cruelty: "In our chosen metaphorics, it was they and not their victims who were the beasts. By treating fellow human beings, beings created in the image of God, like beasts, they have themselves become beasts" (Coetzee and Gutmann 2001, 21). According to Costello, the metaphor of beastliness functions as follows: it is our treatment of others, rather than how we are treated, that makes us beastly. Humans become beasts when we treat one another like beasts. The Jews retained their humanity despite being treated like beasts—it was the Nazis who lost their humanity when they treated others like beasts.

Costello's remarks invite the following questions: Do we become beasts only when we treat other *humans* like beasts? What happens when we treat other *animals* like beasts? Do we become beasts by treating beasts like beasts? Before we answer these questions, it is instructive to examine what it means to be cruel. Paradoxically, because of the purported link between cruelty and animality, cruelty becomes a property that is uniquely human. In *The Beast and the Sovereign*, volume 1, Derrida critically engages Jacques Lacan and Gilles Deleuze, giving us a provocative account of how "bestiality" is understood to be a property exclusive to human.

"Bestiality" has two meanings: "bestiality, characterized *either* as perversion or sexual deviancy, zoophilia that pushes people to make love with beasts or to make love to beasts, *or* as cruelty—this bestiality, this double bestiality (zoophilic or cruel) would also be proper to man" (2009, 69). The two meanings of bestiality are both reserved for man. Bestiality as zoophilia is "proper to man" in the sense that it is applicable only in human sexuality. Only human-animal sexual intercourse is considered bestiality. Animal-animal sexual intercourse—even if it is cross-species—is not bestiality. Consider, for example, the sensational story of "sexually frustrated" Antarctic fur seals "forcing themselves on" penguins on Marion Island. This phenomenon is labelled "misdirected mating" rather than bestiality even though it is a cross-species mating (Chen 2012). Thus "bestiality" is reserved for a specific kind of boundary crossing, namely, the crossing of the human-animal boundary.

If zoophilia is "proper to man" because man is not supposed to mate with beasts, then cruelty is "proper to man" because man is not supposed to be cruel. Indeed, "man [can] be guilty, peccable, precisely because he is capable of good and of perfecting himself, amending himself, capable of confessing and repenting" (Derrida 2009, 105). Humans should, as Seneca contends, transcend their animality and cruelty. We are not supposed to be bloodthirsty like a "wild beast." Bestiality is "proper to man" precisely because we are not meant to be bestial.

If bestiality is "proper to man," then what does it tell us about the animal? Does it make sense to say that a beast is bestial? In *The Animal That Therefore I Am*, Derrida reminds us that because only man can act "beastly," we can talk about the beastliness or the *bêtise* of man, but "there is no sense in speaking of the *bêtise* or bestiality of an animal" (2008, 64).[13] In *The Beast and the Sovereign*, volume 1, Derrida reiterate this point: "It is never said of the beast that it is *bête* [stupid] or *bestial.* The adjective, epithet, attribute *bête*, or "bestial" are never appropriate for animal or beast" (2009, 68). Why is it inappropriate to call a beast "bestial" or *bête*? It comes down to what animals are "supposed" to be. Bestiality is "proper to man" because man is not supposed to be bestial. It makes sense to speak of man's bestiality insofar as zoophilia or cruelty violates our sense of humanity. To put it differently, it makes sense to compare man to beast precisely because man is not beast. By this logic, animals cannot be "bestial" or *bête* because they are supposed to be bestial. It does not make sense to say that beasts are bestial because it is a tautology. Just as animals cannot be naked because nudity is just a fact of their lives,[14] they cannot be beastly because their beastliness is already a given—they are meant to be beastly.[15]

If animals are supposed to act beastly, can we condemn their beastliness? Can animals even be cruel if their violence is a given? According to Derrida, insofar as animals "have no relation to the law," they can neither be "cruel" nor "responsible" (2009, 178). After all, to have a relation to the law one must be "free and sovereign" in the first place (2009, 178). In other words, one must be capable of acting otherwise. Just as guilt presupposes the possibility of innocence, cruelty presupposes the possibility of kindness. If cruelty implies the capacity to act otherwise, then we cannot condemn animals for chasing, devouring, and, in the case of the fur seals, sexually assaulting other creatures because they are presumed to be violent. We do not see these acts as acts of cruelty (even though they do inflict suffering) because we assume that animals cannot be responsible for their violence. As Derrida remarks, "the beast, ignorant of the Law, is not free, neither responsible nor culpable, cannot transgress a Law it does not know, cannot be held to be criminal" (2009, 102).[16]

The relationship between animality and cruelty is paradoxical. The cruelty of the animal is both a necessity and an impossibility. It is a necessity insofar as animals are expected to be "cruel." It is also an impossibility insofar as cruelty is "proper to man"—it does not make sense to speak of the beastliness of a beast. As such, animals simultaneously occupy the two limits of the moral spectrum: they are at once "absolutely innocent" and "absolutely monstrous" (Oliver 2012, 495).[17] Indeed, the two poles of these extremes may simply be two sides of the same coin: it is because of the absolute monstrosity of animal behaviours that they are disenfranchised from the moral community. Nothing can explain these "monstrous" behaviours except their purportedly “cruel” nature. But as soon as they are pushed outside the moral community, they become absolutely innocent. After all, if animals are *by nature* “monstrous,” then they cannot really be held accountable for their “monstrous” acts. If animals are supposed to be monstrous, if it is natural for them to be monstrous, then their monstrosity also serves as a testament to their innocence. The more prevalent these “monstrous” behaviours, the more natural and banal they seem.[18] In sum, just as animals cannot be naked because they are naturally—supposedly—naked, they cannot be cruel because they are naturally—supposedly—cruel.

Given the link between beastliness and cruelty, the portrayal of the Japanese fishermen as heartless dolphin killers is yet another way to animalize them. While Elder, Wolch, and Emel have helpfully distinguished between two mechanisms of dehumanization—presenting “people-as-beasts” and presenting “people-acting-beastly-towards-animals”—this distinction collapses in the end. By highlighting the cruelty of the Japanese fishermen, the documentary presents them *as beasts*, and their beastliness is antithetical to the activists’ humanitarianism. Following Costello, we may say that the Japanese fishermen have become beasts by treating other beasts like beasts. The human-animal opposition reasserts itself at the heart of animal advocacy.

## THE LANGUAGE DIVIDE IN *THE COVE*

Importantly, language difference also plays a role in buttressing the hero-villain opposition in this film. Throughout the documentary, the Japanese fishermen are portrayed as barbaric, dangerous, and inarticulate. Their communication with the activists is, for the most part, confined to poking them with hand-written placards (“Don’t Take Photos”) or yelling at them in Japanese to get out of the killing cove. One fisherman seems to know only two English words, “Private Space,” and he has repeated these two words to the activists long enough that they nicknamed him “Private Space.”

In a scene in which the Japanese police interrogate O’Barry, the language divide is particularly evident. For one thing, the dominant voice in this scene is that of O’Barry—he speaks the loudest and clearest. And while we can also hear the voice of the Japanese translator, the two Japanese police sit silently. The effectiveness of the interrogation is certainly undermined when the person being interrogated appears to be more vocal than the interrogators. Furthermore, the filmmakers have dutifully subtitled all of the translator’s words, presumably because the translator’s accent is nearly incomprehensible. The theatrical effect of the subtitles is significant: while O’Barry’s voice seems unmediated and directly accessible to the audience,[19] the voice of the interrogators is twice removed—first by a translator and then the subtitles. The subtitles in this scene function as cues to the foreignness and otherness of the supposed villain.[20]

Significantly, in the same scene O'Barry brings up the issue of speaking for others. When the translator asks O'Barry about the Oceanic Preservation Society (OPS), O'Barry denies ties with the OPS. He adds that he cannot "speak for" the OPS. Indeed, he tells the police "I can only speak for Ric O'Barry. . . . I cannot speak for anybody except myself." This is a rather intriguing claim, considering that it comes from a dolphin advocate whose mission involves speaking for the dolphin other. O'Barry's response brings us back to the question of whether we can speak for others, be it fellow activists or dolphins, humans or animals.

Contra these foot-stomping, chest-butting Japanese fishermen, the activists recount the plights of the dolphins movingly and eloquently. They offer both sentimental anecdotes and reasoned arguments to persuade us of the horror of the killing cove. In fact, even the dolphins seem to speak. In one scene, director Psihoyos stresses the importance of using hydrophones so that they can hear the dolphins, while in a different scene O'Barry maintains that "[the dolphins] are always trying to communicate with us" and that he can even "read their body language" after spending years living with them. In another scene, O'Barry speaks of the death of Kathy, one of the star dolphins in the show *Flipper*. O'Barry maintains (both in the film and elsewhere) that Kathy "committed suicide" in his arms, and that he sees dolphins committing suicide in Taiji. In a NPR interview, O'Barry reiterates his view as follows: "I see [the suicide] in the cove, in Taiji, almost every day, from September through March. And somebody's standing next [to] me, maybe they don't see it because they can't read the dolphins' body language like I can" (Gross 2009).[21] It is striking that the activists seem to communicate better with the dolphins than the Japanese fisherman. The language barrier, it appears, bifurcates the dolphin rescuers and the dolphin killers, heroes and villains, good and evil.

Although the language barrier arises rather innocuously as the activists and the fishermen do not share a language, it becomes complicit in the mechanism of dehumanization. The linguistic divide operates in a way similar to the cruelty-kindness divide that we have seen earlier. As we know, throughout the history of Western thought the human is posited as the speaking animal, the animal with *logos*. As such, the human-animal divide is simultaneously a linguistic divide. Given this philosophical backdrop, the portrayal of the fishermen as violent and inarticulate is yet another way to animalize them. The language barrier is employed (even if unintentionally) to reinscribe the human-animal linguistic divide: whereas the eloquent activists are humanitarians, the inarticulate fishermen are cruel dolphin murderers.

My analysis of *The Cove* reveals that the dolphin activists unwittingly fall back on the human-animal opposition in their advocacy for the animal. By portraying the Japanese as violent and inarticulate, the filmmakers highlight the two elements that are commonly attributed to the animal. This mechanism of animalization at best marginalizes the racial other as the moral underclass and at worst disenfranchizes the racial other from the moral community. Animalization goes hand in hand with racialization and marginalization.

## REVISITING THE CRUELTY-BEASTLINESS LINK

At this point, I want to return to the discussion of cruelty that I introduced in chapter 1. Recall that in her provocative essay, "How to Do Things with Animals," Karen Raber problematizes the wholesale rejection of cruelty that posthumanist animal advocates often endorse. She faults posthumanists for harbouring a "fantasy of the post-cruel"—a "desire to eradicate human abuse of animals in all its forms" (2008, 106). One illustrative example of this fantasy of the "post-cruel" is the vegan pet food movement. In their attempt to resolve the moral dissonance of keeping a vegan diet while buying meat for their pets, some vegan animal caretakers have begun feeding their pets a vegan diet in recent years. In a CNN interview, a vegan animal caretaker remarks, "the fact we have a vegan option is a double-benefit because it means our dog can live with the same ideology" (Grinberg 2011).

According to Raber, the fantasy that we can be cruelty-free is fundamentally at odds with the project of the posthumanist. Insofar as cruelty is part of what it means to be an animal, the very rejection of cruelty (towards animals) is a disavowal of our animality. By rejecting cruelty and their own animality, animal advocates remain faithful to the very anthropocentrism that they seek to challenge. At the end of her paper, Raber wonders whether it is "possible that we *need* human exceptionalism" in animal ethics (2008, 101). Her point is that we do need this exceptionalism in order to speak of our moral responsibility to animals; we do need the "violence" of this human exceptionalism so that we can curb the worse violence that we inflict on animals.

However, even if human exceptionalism is inevitable, we should still reconsider the problematic link between cruelty and beastliness. The "conundrum" of the post-cruelty stance exists only if we continue to link cruelty with animality. Rejecting cruelty means rejecting animality only if we continue to assume that beastliness is a given of the beast or that cruelty is integral to animality. But surely not all animals are predators. In fact, *most* animals are not predators. Furthermore, couldn't predators also love? Isn't the cat that chases and torments the squirrel the very same cat that plays with us? Isn't love or kinship also possible for the most ferocious predator? My point is not to romanticize animals and the suffering that they are capable of inflicting, but to challenge the link between beastliness and cruelty. It seems that animals—and dare I say *like humans*—are at once cruel and tender, dangerous and gentle, hostile and affectionate. Even if we must embrace human exceptionalism, we should still undo the insidious link between beastliness and cruelty—a link that has served racism and colonialism. Breaking this link helps animal activists resist animalizing the other in their animal advocacy. After all, if cruelty were not linked to the beasts, then the strategy of portraying others as "acting *beastly* towards animals" would lose its rhetorical force.

## ANIMAL VIOLENCE AND NATIONAL IDENTITY

Thus far, I have argued that the human-animal divide resurfaces in animal advocacy when we continue to link humaneness with humanity, cruelty with animality. By insisting that animal violence is "*in*humane," by calling animal abusers "subhuman," we animalize those who act cruelly towards animals. In what follows, I will consider the question of identity, specifically the ways in which animal cruelty is employed to articulate cultural, as well as national, identities.

Recall Deckha identifies the double standard at work in early anticruelty legislation: anticruelty laws selectively target animal abuse associated with the lower classes while turning a blind eye to animal abuse such as fox hunting, which is typically practiced by the upper classes. The legacy of this double standard can be found in contemporary anticruelty law, as it "continues to selectively target minoritized practices as 'cruel' while immunizing the vast range of normalized acts of violence against animals" (2013, 526). The Michael Vick dogfighting scandal is a case in point—the public outrage stirred by the scandal is unmatched by outrage over the callous treatment of animals in factory farms. While dogfighting (which is typically associated with African American) is illegal, horse racing is considered a legitimate "sport." Indeed, the annual Kentucky Derby, where mint juleps are served and big, ornate hats are worn, is considered a cultural event rather than animal cruelty.

Identifying the double standard is not to suggest that blood sports and dogfighting should be tolerated just like fox hunting and horse racing—animal abuse is bad for animals regardless of how it is done and who does it. However, bringing to light the double standard helps us identify whose interest animal protection is actually serving, and who (in addition to the cats and dogs) may benefit from such a selective animal advocacy. As we have seen earlier, Deckha argues that the apparent hypocrisy of targeting a specific kind of animal abuse served a larger and more insidious purpose: during the heyday of colonial expansion, it enabled the empire to establish itself as a moral authority without compromising the interests of the upper class. This moral authority (on how animals should be treated), in turn, articulated the British national identity and superiority. Whereas military might was used to conquer lands, moral authority was used to *justify* dominion.

Deckha's insight is highly instructive in the context of American politics. Whereas the selective animal advocacy in the past has built moral and national superiority, the selective animal advocacy in the present has furthered anti-immigration agenda. For example, Carrying Capacity Network (CCN), an anti-immigration group, explicitly ties immigrants to animal abuse. According to an article on their website, immigrants are responsible for "importing third-world cruelty" such as horse tripping, animal sacrifice, and cockfighting, to America (2005). According to CCN, these animal practices are indicative of the sadism found in these "third-world cultures." And since "stopping such culturally sanctioned practices is very difficult, the only solution lies in a combination of publicity and, ultimately, stopping mass immigration" ("Importing Violence" 2005). This rhetoric is striking: immigrants, who are *by culture* cruel to animals, are importing violence to America, and the only sensible solution to this problem is to keep them out of the country. (It is as if violence did not exist in America until these "third-world" immigrants "imported" it.)

The connection between immigrants and animal abuse is made not only by extreme anti-immigration groups like CCN, but also by politicians. Representative Elton Gallegly (R-California) has linked illegal immigration with animal abuse in a 2009 op-ed. (Gallegly, an ardent anti-immigration apologist, is also famous for passing the anticruelty bill that outlaws animal-crush fetish videos.) In this op-ed, Gallegly makes the point that "animal abusers are more likely to engage in other illegal acts" (Gallegly 2009). These illegal activities are mostly connected to cockfighting and drug trafficking by gangs and drug cartels. In Phoenix, Arizona, however, one crime syndicate not only raised roosters for cockfighting, but also "preyed on illegal immigrants, subjecting them to kidnappings, home invasions and armed robberies" (Gallegly 2009). Both the roosters and the illegal immigrants are victims of the crime syndicate.

Illegal immigrants and animal abuse are linked not just in "exotic" practices such as cockfighting. In early 2014, NBC aired a video revealing the cruel treatment of pigs at an Oklahoma pig farm. The pigs in the video were pulled, kicked, shaken, and thrown, while piglets were killed by being smashed onto the floor. After the video aired, Tyson Foods changed its "pig care policies" and cut ties with the supplier shown in the video. The owner of the farm was purportedly "stunned" by the mistreatment of the animals and immediately fired the employees seen on the video.

While Tyson Foods changed its "pig care guidelines" to end "blunt force euthanasia" (which is an oxymoron, given that euthanasia means "good death"), it does not address the *underlying* factors that contributed to animal abuse on the pig farm. Indeed, the responses of both Tyson Foods and the farm owner suggest that we should direct our blame to the specific abusers rather than the system. Cutting ties with the one supplier and firing specific offending employees give the false impression that the problem of animal abuse is isolated rather than systemic and structural. The owner of the pig farm predictably claimed that his farm used "approved methods of euthanasia on animals," implying that the fault really belonged to the few bad apples in the farm, who he fired immediately. The reactions of Tyson Foods and the owner of the pig farm presuppose the following: the issue of animal abuse has nothing to do with the farm management, given that they are supposedly rule abiding; it has nothing to do with the systematic, large-scale production of meat, given that the new animal care guideline is supposed to take care of the problem; and it certainly has nothing to do with the practice of meat eating, given that even the *consumers* are outraged by the abuse. If none of these institutions are at fault, then the real culprits have to be the callous, sadistic workers who wilfully disregarded the "approved methods of euthanasia" and purposefully inflicted pain on the pigs. Without addressing the underlying factors that make animal abuse a systemic and structural issue in these factory farms, animal violence becomes the abuser's individual responsibility.

So who are these supposedly callous, sadistic animal abusers? While it is not easy to discern from the video, it is well known that many Latino immigrants work for slaughterhouses. The meatpacking industry has changed dramatically since the 1970s: whereas in the early 1970s it was one of the highest-paid industrial jobs in the United States with the lowest turnover rate, the reverse is true on both counts nowadays (Schlosser 2001). This is in part due to the meatpacking industry's active recruitment of Latino immigrants, many of whom work without documentation. Immigrants are more willing to accept the low wages, long hours, as well as the unsavoury jobs in the slaughterhouse. As big corporations move slaughterhouses to rural areas, these rural places see an influx of immigrants, who are often scapegoated for the increased crime rates in slaughterhouse communities (Fitzgerald, Kalof, and Dietz 2009, 4). As such, even before the NBC video of the pig farm abuse, there had already been a prejudiced public perception linking immigrants, slaughterhouses, and criminality. In the end, the link between animal abuse and immigrants has long been established and the video only strengthened this existing link.

So what does this purported connection between immigrants and animal abuse amount to? Recall again Deckha's argument: by selectively targeting animal violence associated with the lower class, the colonial British Empire could export moral superiority without compromising the upper class' interests. As we have seen, anti-immigrant groups such as CCN use the rhetoric of "third-world cruelty" to advocate tighter border control and immigration restriction. By selectively targeting minorities and immigrants, CCN establishes moral authority without jeopardizing the dominant group's interest in consuming meat. As with the case of the British Empire, this moral authority is supposed to establish national identity and superiority—without guns and cannons.

But the nationalistic superiority that we see in colonial British Empire is reincarnated in anti-immigration discourse in a different way. While the British Empire established its national and moral superiority by civilizing the "savages," anti-immigration rhetoric articulates the American identity by insisting that immigrants are *incorrigible*. According to CCN, animal violence is an integral part of the "third-world's cultures." Given that there is no way to stop culturally ingrained animal abuse, the only sensible action is to keep immigrants and their animal violence out of the country. Now this may appear to be some extremist rhetoric made by a fanatic anti-immigration group, but as we will soon see, American immigration law is working in tandem with the idea that immigrants are incorrigible, thus undeserving of moral education. It is by preventing immigrants from "importing" animal violence that the real American identity can be safeguarded—a real American knows which animals to love and which to eat. A real American also knows how to *use* an animal in the right way: training roosters to fight is cruel, but making gourmet chicken dishes is not.

The apparent hypocrisy of selective animal protection in the United States serves a more insidious purpose of immigration control. But this is not just a matter of rhetoric or propaganda—animal abuse can indeed be used as a reason for deportation. An immigration official can either downgrade or deport a green card applicant if the applicant is convicted of crimes of "moral turpitude," which include animal fighting. It should be noted that the concept of "moral turpitude" is rather broad; in the "Matter of Agustin ORTEGA-LOPEZ,"[22] it refers to conduct which is "inherently base, vile, or depraved, and contrary to the accepted rules of morality and the duties owed between persons or to society in general" (Malphrus 2013, 100). In the same decision, the board states that the fact that dogfighting and cockfighting are illegal in all fifty states and the District of Columbia shows that "we, *as a society*, find animal fighting morally reprehensible, and thus morally turpitudinous" (2013, 103; emphasis mine). By making animal fighting a *deportable* crime, the law implicitly endorses the following idea: the proper way to deal with animal cruelty committed by immigrants is not moral education (as the colonizers would have done), but exclusion and banishment. As such, the American immigration law is working in tandem with the seemingly extreme claim made by the CCN that immigrants are morally incorrigible.

## MARGINALIZING OTHERS IN THE NAME OF CULTURAL DIVERSITY

In closing, I will return to a question that I asked earlier: How do we voice our dissent about the oppressive features of traditional cultural practices in a way that does not reinscribe both colonialism/racism and human exceptionalism? Does respecting cultural diversity mean exempting non-Western cultures from the charge of animal abuse? In "Is Multiculturalism Bad for Animals?" Paula Casal argues that respecting cultural diversity does not justify more animal suffering (2003). For one thing, just because something is a tradition does not mean it should be preserved, for the obvious reason that not all traditions are good traditions. The demand to respect other cultures does not mean giving *carte blanche* to traditions that continue to inflict pain on animals. For Casal, the fact that some cultural practices have been targeted but not others (say the consumption of dog meat is reprimanded but not the consumption of chicken or pork) does not mean that neither should be called out. As Casal puts it, "the fact that many murderous dictators have escaped the punishment they deserve is no good reason not to try other dictators whenever feasible—let alone to create legal exemptions for other individuals engaging in comparable activities" (2003, 19). Similarly, the fact that animal advocacy has been appropriated by racist and xenophobic discourse does not entail that racial or cultural minorities should be exempted. In fact, such exemptions will lead to a rather absurd position in which we can only speak for animals that are abused by those in the position of power, but not animals that are abused by racial or cultural minorities.

In *Zoopolis*, Donaldson and Kymlicka offer an illuminating discussion of the tension between respecting cultural diversity and universalizing animal rights (2011, 44–49). They remind us that a similar debate takes place in the human rights discourse: some have challenged the universality of human rights on the grounds that the concept of rights is primarily Western, so to claim universality is to *impose* Eurocentric ideals on non-Western cultures (2011, 45). But as Donaldson and Kymlicka argue, the claim that the concept of human rights is Western involves a problematic assumption: human rights are specific to the cultural DNA of the West. It is somehow "natural" for the West to evolve and flourish with the concept of human rights. Eastern cultures, on the other hand, purportedly lack the appropriate cultural DNA to accommodate the concept of human rights (2011, 45–46). Donaldson and Kymlicka reject the "cultural DNA" presumption; they maintain that a culture comes to endorse human rights not because it is somehow "destined" to do so, but through careful reflection and selection from various moral sources. For them, the same reasoning applies to *animal* rights (2011, 46–47).

> No society is predetermined to embrace ART [animal rights theory], but nor is any society predetermined to reject it. Every society contains a diversity of moral sources on the status of animals, some of which lead comfortably in the direction of ART, others of which do not, and it is up to all of us to judge which of these moral sources we find compelling. (2011, 46)

To say that animal rights are *imposed* upon non-Western cultures is to assume that it is somehow "unnatural" for non-Western cultures to adopt animal rights. It is as if non-Western cultures lack the appropriate cultural DNA to accommodate animal rights. But perhaps there is a more problematic assumption here: it is not just that non-Western cultures fail to accommodate animal rights, but that animal violence is an integral part of these cultures. More specifically, the assumption is that the cultural identity of these non-Western groups is tied to practices that involve animal violence (Spaniards to bullfighting, Latinos to cockfighting, etc.), and that we fail to "respect" their cultural identity when we call for the end of these practices. The assumption that a culture is tied to practices of animal abuse is eerily similar to the rhetoric of the CCN. Recall that the CCN maintains that animal violence (e.g., cockfighting) is ingrained in the "third-world's cultures," hence it is both impossible and unnecessary to correct culturally ingrained animal abuse. The only sensible action is to keep immigrants and their animal violence out of the country. These "third-world cultures" are essentially "exempt" from embracing animal rights insofar as they are supposedly morally incorrigible. The effort to promote cultural diversity may actually invoke problematic assumptions about ingrained violence that we see in racist and xenophobic discourse.

This brings us back to the notion that beastliness is a given of the beast. The presumption that animal violence is culturally ingrained is yet another expression of the dubious link between cruelty and animality—just as cruelty is (purportedly) what it means to be an animal, participating in cockfighting is what it means to be a Latino immigrant; just as it makes no sense to talk about the *bête* of an animal, it is futile to educate the morally incorrigible animal abusers. Furthermore, the idea that non-Western cultures should be exempted from being held accountable for animal abuse parallels the idea that animals cannot be faulted for their violence. The assumption is that one cannot be held accountable for being cruel if one is *supposed* to be cruel—whether by nature or by culture. As such, exempting cultural and racial minorities from charges of animal abuse may reinforce the idea that animal violence is ingrained in these non-Western cultures. The very effort to respect cultural diversity may inadvertently essentialize non-Western groups as animal abusers. Once again, the link between cruelty and animality is at the service of racist discourse.

## NOTES

1. Deckha offers a helpful account (2012) of the role of race (and culture) in posthumanist feminist writing. She argues that despite its sensitivity to intersectionality, feminists continue to privilege gender discourse when they examine animal oppression. She argues for the urgent need to put race/culture front and centre in posthumanist feminist discourse.

2. For example, Claire Jean Kim and Carla Freccero edited a special issue on "Species/Race/Sex" in *American Quarterly* that examines how species, race, and sex intertwine (2013). Kim's *Dangerous Crossings* examines the way animal welfare clashes with multiculturalism (2015).

3. Megan Glick also examines Michael Vick's dogfighting scandal in her essay "Animal Instincts: Race, Criminality, and the Reversal of the 'Human'" (2013).

4. In her recent book, Lisa Guenther (2013) offers a provocative analysis of solitary confinement as a means of dehumanization. She complicates the idea of animalization as dehumanization by arguing that we often fail to treat animals *as animals*.

5. See Roberts (2008, 17–20, 81–83).

6. In his infamous diatribe against a Mexican teenager who was accused of incest with his sister, Judge Gerald Chargin compares the teenager (as well as Mexican people more generally) to an animal: "You are just an animal. You are lower than an animal. Even animals don't do that. You are pretty low . . . Mexican people, after 13 years of age, it's perfectly all right to go out and act like an animal" (Haney-López 2003, 84).

7. "Animal practice" refers to what we do to the animal.

8. "Chinese region" refers to a region that is officially a part of the People's Republic of China (PRC). While the consumption of cats and dogs is also prohibited in Taiwan—the Republic of China—I leave it as it is considered a sovereign state.

9. Vizcarrondo (2009).

10. This scene is found in a DVD version of *The Cove*.

11. I do not mean to suggest that O'Barry has *falsely* implicated the "Japanese mafia." Their actual involvement is beside the point. My analysis is mainly concerned with the *portrayal* of the Japanese fishermen as villains in the documentary.

12. This point is implicit in Elder, Wolch, and Emel's account, but they do not articulate it in their paper.

13. Derrida is playing on the etymological connection between "the beast" (*la bête*) and "stupidity" (*bêtise*) in French. So only man can be "stupid" or "beastly."

14. See Derrida (2008, 4–5).

15. Also, if being beastly is what the beasts do, then is being humane what humans do? Put differently, if it makes little sense to call a beast "beastly," does it make sense to laud a human for being "humane"?

16. That cruelty is a uniquely human trait also stems from the idea that cruelty must be accompanied by the *intent* to harm. One can be harmed by a natural disaster such as a tornado or an unfortunate automobile accident. But it would be a mistake to speak of the "cruelty" of a tornado (except as a metaphor in which the tornado is personified as an *intending* agent) or the "cruelty" of a novice driver. We are unwilling to describe these harm-inflicting incidents as "cruel" because we conceive of cruelty as wilful actions that intend to harm, and it is precisely this intention to harm that renders such actions morally blameworthy.

17. In *The Beast and the Sovereign*, volume 1, Derrida reminds us that animals are presumed to be "innocent" (2009, 124).

18. For an insightful analysis on racism, accountability, and animality, see Peterson (2010). In this essay, Peterson analyses the character of Bigger Thomas in Richard Wright's 1940 novel *Native Son*. He argues that Bigger is portrayed as both an animal and a human. Bigger is portrayed as an animal insofar as his crime seems "natural" and that his actions are "instinctive" (2010, 165). But insofar as Bigger is held accountable for his crime, "the prosecution inadvertently confers a certain humanity on him" (2010, 164).

19. The target audience of this film is, of course, the English-speaking Westerner. Both the film and the DVD of this film were first released the United States.

20. My point is not that the filmmakers have *unnecessarily* put on the subtitles *in order to* highlight the foreignness of the Japanese—this editorial decision need not be motivated by any agenda. I am merely pointing out the *theatrical effect* of this specific scene: O'Barry's voice is unmediated, whereas the voice of the Japanese interrogators is twice removed from the audience: first by the translator, then by the subtitles.

21. In the documentary, O'Barry also explains that dolphins do not breath automatically, and that "every breath they take is a conscious effort." Thus, "They can end their lives whenever *life's become too unbearable*" (my emphasis).

22. This is a case concerning the deportation of Agustin Ortega-Lopez. The decision is written by Grant Malphrus, one of the board members.

# Epilogue

In closing, I would like to return to the significance of the human-animal language divide in animal ethics. As we have learned from the history of philosophy, our identity as human is often constructed by pitting ourselves against the animal: we are said to be unique, we are said to be different from other animals by virtue of our capacity to speak. Language has long been the quality that defines our humanity. I began this book by asking whether humans, insofar as we can speak, should take up the responsibility of speaking for animals. Throughout the book I demonstrated the necessity of speaking for animals, and to speak for them responsibly. In chapters 1 and 4, I argued that representation, translation, and interpretation are inevitable in animal advocacy. The injunction to let animals speak for themselves betrays a problematic assumption that there is a pure, uncontaminated animal voice free from human biases. Insofar as representation is inevitable, animal advocates should strive to speak well for animals, to speak more responsibly on their behalf. In chapter 3, I argued that one way to maintain the "otherness" of the animal is to acknowledge the asymmetry of moral responsibility. That is, we should speak for animals even if they do not reciprocate; we are responsible to animals even if they refuse to respond. And finally in chapter 5, I argued that in order to speak for animals responsibly, we must interrogate the ways in which racism and speciesism intersect. It is important to be critical of the rhetoric that we employ so that animal advocacy does not become another means of oppression. As such, my overall argument in this book is that we cannot shrink from our responsibility to speak for animals, even if speaking for animals carries ideological baggage that reinscribes human exceptionalism.

Even though I have argued that we need to give voice to the (politically) voiceless animals by speaking for them, I would like to end my book on a rather different note. Let us return briefly to Freud and his beloved dogs. As we saw in chapter 2, dogs served as various substitutes for the father of psychoanalysis. They were surrogate of a protective father, replacement of a lost child, and even stand-in for Freud's digestive system. Indeed, Freud's dogs served him both personally and professionally, as the dogs made their way into the therapy room as well. According to Stanley Coren, the presence of a dog seems to help the patients speak:

> This difference was most marked when Freud was dealing with children and adolescents, who seemed more willing to talk openly (especially about painful issues) when the dog was in the room. . . . In addition, during psychoanalysis, when the patient is getting near to uncovering the source of his or her problem there is often a resistance phase. . . . Freud's impression was that the resistance was so much less vigorous when the dog was in the room. (Coren and Bartlett 2002, 139–40)

Freud's observation that patients are more willing to speak in the presence of a dog is illuminating when we consider the ways service animals are trained to assist our lives nowadays. In a recently developed program called "Tail Waggin' Tutors," children practice their reading skills with their dog "tutors" by reading books out loud to them. The program is supposed to help children build their self-esteem: unlike human tutors, canine tutors do not "judge" (or care) when a child says a word wrong (Shammas 2015). They help children improve their reading skills not by correcting them, but by giving them a safe, non-judgmental speaking environment.

Another form of animal-assisted learning that is gaining popularity in recent years is the use of service dogs in autism therapy. As we saw in chapter 3, Greenspan and Shanker offer a developmental and evolutionary account of language formation (2004). Their model of language formation is motivated in part by their research with autistic children. As they point out, autistic children often have difficulty developing both language and social skills. Their hypothesis is that the crucial developmental stage of emotional signalling is disrupted by certain biological factors in children with autism. Without mastering the skill of negotiating emotions with their caretakers, autistic children have difficulty attaching emotionally rich images to words, which hinders language development. As such, their language impediment is a consequence of their social-emotional impediment (2004). Fortunately, despite the fact that autistic children often have difficulty bonding with their siblings and parents, many are able to bond with their therapy dogs. In one case study, an autistic child demonstrates a marked improvement in social competence, as well as spontaneity in verbal communication, in the presence of her therapy dog (Solomon 2010). Therapy dogs are now widely employed to help autistic children gain confidence and facilitate social acceptance.

But trained dogs are not merely helping children with special learning needs. In the past few years, dogs trained by assistance dog organizations have served as "testimony enablers" in numerous courthouses. These courthouse facility dogs are allowed in some courtmans to provide comfort for vulnerable or traumatized witnesses, such as children who were sexually abused. For example, in 2010 a courthouse facility dog (Rosie) provided emotional support for a fifteen-year-old who was raped and impregnated by her father: "[Rosie] sat by the teenager's feet. At particularly bad moments, she leaned in [to encourage the witness]" (Glaberson 2011). Just as Freud's dogs helped the patients confront their own demons by overcoming the "resistance phase," Rosie helped the teenage girl confront her father. And just as Freud's dogs helped the patients speak about their issues, Rosie helped a reluctant and frightened teenager to speak for herself and against her abuser. Advocates for the "dog-in-court-cause" argue that just as the "Support Persons Case Law" allows a support person to "increase some children's capacity to testify and enhance the child's direct and cross-examination," a specially trained dog should be allowed to serve the same purpose (O'Neill-Stephens, 2010).[1] Courthouse dogs provide the emotional support that young witnesses need to testify in court, often against their abusers. Indeed, more recently courthouse dog service has even been made part of public policy. In 2015, Arkansas passed House Bill 1855, known as "The Courthouse Dog Child Witness Support Act," allowing child witnesses under eighteen to be accompanied by specially trained facility dogs when testifying in criminal trials. As witness testimony is an essential part of justice, it is difficult to exaggerate the importance of the service of these courthouse facility dogs.

In all three examples above, trained dogs offer a common service—helping children and the vulnerable speak. By bonding with the canine other, schoolchildren practice their reading skills, autistic children develop competence in spontaneous communication and social interaction, and child witnesses wield the power to testify in court. In an important sense, these dogs are giving voice to these children. In light of the popular dictum of "giving voice to the voiceless" in animal advocacy, these trained animals remind us that animals have already been giving voice to the most voiceless, most vulnerable among us. I argued in chapter 1 that the injunction to let animals speak for themselves is a fantasy because we cannot avoid representing animals in animal advocacy. These dogs show that they can accomplish what remains a fantasy for us: they help children speak not by representing or speaking for them, but by giving them companionship, confidence, and love.

Significantly, it is by establishing a relationship with the animal that these children gain the confidence to speak. In the case of the courthouse facility dog, the teenage girl found her voice in court by bonding with—rather than identifying herself against—Rosie the dog. Rosie rehumanizes the teenage girl by giving her the courage to speak. This flies in the face of the philosophical cliché that defines humanity by setting it against the non-speaking animal. These animals invite us to rethink and reconfigure the triangular relationship between language, humans, and animals. In the end, perhaps we as humans are constituted as speaking beings not by setting ourselves in opposition to animals, but rather by exercising our capacity to *relate to* them. We become the most human, most capable of speaking, when we bond with our animal others.

## NOTE

1. There are now five appellate court decisions that specifically address the use of courthouse facility dog to assist vulnerable witnesses, hence advocates no longer have to rely on "Support Person Case Law." (Thanks to Ellen O'Neill-Stephens for the pointer.)

# Bibliography


Abram, David. 1996. *The Spell of the Sensuous: Perception and Language in a More-than-Human World.* New York: Random House.

Adams, Carol J. 1990. *The Sexual Politics of Meat: A Feminist-Vegetarian Critical Theory.* New York: Continuum.

Adams, Carol J. 1994. *Neither Man nor Beast: Feminism and the Defense of Animals.* New York: Continuum.

Aho, Kevin. 2007. "Logos and the Poverty of Animals: Rethinking Heidegger's Humanism." *The New Yearbook for Phenomenology and Phenomenological Philosophy*: 109–26.

Alcoff, Linda. 1991. "The Problem of Speaking for Others." *Cultural Critique* (20): 5–12.

Alexander, Brian. 2009. "Amid Swine Flu Outbreak, Racism Goes Viral." *Msnbc.com*, May 1. http://www.nbcnews.com/id/30467300/ns/health-cold_and_flu/t/amid-swine-flu-outbreak-racism-goes-viral/#.VUTDi9NViko.

Andersen, Wayne V. 2001. *Freud, Leonardo da Vinci, and the Vulture's Tail: A Refreshing Look at Leonardo's Sexuality.* New York: Other Press.

Andrew-Gee, Eric. 2012. "Stop Saving These Animals." *The New Republic*, January 9.

Aristotle. 1995. *The Politics.* Translated by Ernest Barker. Kindle edition. Oxford: Oxford University Press.

Armitage, Simon. 2010. "Morrissey Interview: Big Mouth Strikes Again." *Guardian*, September 3.

Armstrong, Philip. 2008. *What Animals Mean in the Fiction of Modernity.* London: Routledge.

Aronson, Stanley. 2010. "Racism and the Threat of Influenza." *Medicine and Health Rhode Island* 93 (1): 3.

Bailey, Cathryn. 2007a. "'Africa Begins at the Pyrenees': Moral Outrage, Hypocrisy, and the Spanish Bullfight." *Ethics and the Environment* 12 (1): 23–38. doi:10.2979/ETE.2007.12.1.23.

Bailey, Cathryn. 2007b. "We Are What We Eat: Feminist Vegetarianism and the Reproduction of Racial Identity." *Hypatia: A Journal of Feminist Philosophy* 22 (2): 39–59. doi:10.2979/HYP.2007.22.2.39.

Bartkowski, Frances. 2008. *Kissing Cousins: A New Kinship Bestiary.* New York: Columbia University Press.

Bentham, Jeremy, J. H. Burns, and H. L. A. Hart. 1996. *An Introduction to the Principles of Morals and Legislation.* Oxford: Clarendon Press.

Bourdieu, Pierre, and John B. Thompson. 1991. *Language and Symbolic Power.* Cambridge, MA: Harvard University Press.

Brown, Wendy. 1995. *States of Injury: Power and Freedom in Late Modernity.* Princeton, NJ: Princeton University Press.

Butler, Judith. 2005. *Giving an Account of Oneself*. New York: Fordham University Press.

Calarco, Matthew. 2004. "Heidegger's Zoontology." In *Animal Philosophy: Essential Readings in Continental Thought*, edited by Matthew Calarco and Peter Atterton, 18–30. London: Continuum.

Calarco, Matthew. 2005. "'Another Insistence of Man': Prolegomena to the Question of the Animal in Derrida's Reading of Heidegger." *Human Studies* 28 (3): 317–34. doi:10.1007/s10746-005-7413-y.

Carbone, Larry. 2004. *What Animals Want: Expertise and Advocacy in Laboratory Animal Welfare Policy*. New York: Oxford University Press.

Casal, Paula. 2003. "Is Multiculturalism Bad for Animals?" *Journal of Political Philosophy* 11 (1): 1–22. doi:10.1111/1467-9760.00164.

Cavalieri, Paola. 2001. *The Animal Question: Why Nonhuman Animals Deserve Human Rights*. Oxford: Oxford University Press.

Cave, George S. 1982. "Animals, Heidegger, and the Right to Life." *Environmental Ethics* 4 (3): 249–54. doi:10.5840/enviroethics19824329.

Chen, Ingfei. 2012. "In Nature, Fatal Attractions Can Be Part of Life." *New York Times*, November 26. http://www.nytimes.com/2012/11/27/science/in-nature-fatal-attractions-can-be-part-of-life.html?_r=0.

Coetzee, J. M., and Amy Gutmann. 1999. *The Lives of Animals*. Princeton, NJ: Princeton University Press.

Cohen, Carl, and Tom Regan. 2001. *The Animal Rights Debate*. Lanham, MD: Rowman and Littlefield.

Colony, Tracy. 2007. "Before the Abyss: Agamben on Heidegger and the Living." *Continental Philosophy Review* 40 (1): 1–16. doi:10.1007/s11007-007-9047-4.

Coren, Stanley, and Andy Bartlett. 2002. *The Pawprints of History: Dogs and the Course of Human Events*. New York: Free Press.

Curtin, Deane. 1991. "Toward an Ecological Ethic of Care." *Hypatia* 6 (1): 60–74. doi:10.1111/j.1527-2001.1991.tb00209.x.

Curtin, Deane. 2014. "Compassion and Being Human." In *Ecofeminism: Feminist Intersections with Other Animals and the Earth*, edited by Carol J. Adams and Lori Gruen, 39-57. Kindle edition. New York: Bloomsbury Academic.

Deckha, Maneesha. 2012. "Toward a Postcolonial, Posthumanist Feminist Theory: Centralizing Race and Culture in Feminist Work on Nonhuman Animals." *Hypatia* 27 (3): 527–45. doi:10.1111/j.1527-2001.2012.01290.x.

Deckha, Maneesha. 2013. "Welfarist and Imperial: The Contributions of Anticruelty Laws to Civilizational Discourse." *American Quarterly* 65 (3): 515–48. doi:10.1353/aq.2013.0033.

DeGrazia, David. 1994. "Wittgenstein and the Mental Life of Animals." *History of Philosophy Quarterly* 11 (1): 121–37.

Deleuze, Gilles, and Félix Guattari. 1987. *A Thousand Plateaus: Capitalism and Schizophrenia*. Minneapolis: University of Minnesota Press.

Derrida, Jacques. 1987. "Geschlecht II: Heidegger's Hand." In *Deconstruction and Philosophy: The Texts of Jacques Derrida*, edited by John Sallis, 161–196. Chicago: University of Chicago Press.

Derrida, Jacques. 2000. "Hospitality." *Angelaki: Journal of the Theoretical Humanities* 5 (3): 3–18.

Derrida, Jacques. 2008. *The Animal That Therefore I Am*. New York: Fordham University Press.

Derrida, Jacques. 2009. *The Beast and the Sovereign*. Vol. 1. Chicago: University of Chicago Press.

Diamond, Cora. 1978. "Eating Meat and Eating People." *Philosophy* 53 (206): 465–79. doi:10.1017/S0031819100026334.

Diamond, Cora. 2001. "Injustice and Animals." In *Slow Cures and Bad Philosophers: Essays on Wittgenstein, Medicine, and Bioethics*, edited by Elliott Carl, 118–148. Durham, NC: Duke University Press.

Diamond, Cora. 2003. "The Difficulty of Reality and the Difficulty of Philosophy." *Partial Answers: Journal of Literature and the History of Ideas* 1 (2): 1–26. doi:10.1353/pan.0.0090.

Dombrowski, Daniel A. 1997. *Babies and Beasts: The Argument from Marginal Cases*. Urbana: University of Illinois Press.

Donaldson, Sue, and Will Kymlicka. 2011. *Zoopolis: A Political Theory of Animal Rights*. Kindle edition. Oxford: Oxford University Press.

Donovan, Josephine. 1990. "Animal Rights and Feminist Theory." *Signs: Journal of Women in Culture and Society* 15 (2): 350–75. doi:10.1086/494588.

Donovan, Josephine. 2007. "Animal Rights and Feminist Theory." In *The Feminist Care Tradition in Animal Ethics: A Reader*, edited by Josephine Donovan and Carol J. Adams, 58–86. New York: Columbia University Press.

Donovan, Josephine, and Carol J. Adams. 2007. *The Feminist Care Tradition in Animal Ethics: A Reader*. New York: Columbia University Press.

Dunayer, Joan. 2001. *Animal Equality: Language and Liberation*. Derwood, MD: Ryce Pub.

Durantaye, Leland De La. 2005. "Vladimir Nabokov and Sigmund Freud, or a Particular Problem." *American Imago* 62 (1): 59–73. doi:10.1353/aim.2005.0011.

Elden, Stuart. 2006. "Heidegger's Animals." *Continental Philosophy Review* 39: 273–91.

Elder, Glen, Jennifer Wolch, and Jody Emel. 1998. "Le Pratique Sauvage: Race, Place, and the Human-Animal Divide." In *Animal Geographies: Place, Politics, and Identity in the Nature-culture Borderlands*, edited by Jennifer Wolch and Jody Emel, 72–90. London: Verso.

Elliott, Carl. 2001. *Slow Cures and Bad Philosophers: Essays on Wittgenstein, Medicine, and Bioethics*. Durham, NC: Duke University Press.

"Empower, v." 2015. *OED Online*. Oxford University Press. http://www.oed.com/view/Entry/61399?redirectedFrom=empower#eid.

Fellenz, Marc R. 2007. *The Moral Menagerie: Philosophy and Animal Rights*. Urbana: University of Illinois Press.

Fitzgerald, A. J., L. Kalof, and T. Dietz. 2009. "Slaughterhouses and Increased Crime Rates: An Empirical Analysis of the Spillover from 'The Jungle' into the Surrounding Community." *Organization and Environment* 22 (2): 158–84. doi:10.1177/1086026609338164.

Foltz, Bruce. 1993. "Heidegger, Ethics, and Animals." *Between the Species: A Journal of Ethics* 9 (2): 84–89.

Francione, Gary. 2000. *Introduction to Animal Rights: Your Child or the Dog?* Philadelphia: Temple University Press.

Francione, Gary. 2009. "Animal Welfare and the Moral Value of Nonhuman Animals." *Law, Culture and the Humanities* 6 (1): 1–13.

Francione, Gary. 2011. "No Moral Distinction between Furs and Leathers." *New York Times*, September 26. http://www.nytimes.com/roomfordebate/2011/09/25/ban-fur-then-why-not-leather/no-moral-distinction-between-furs-and-leathers.

Francis, Leslie Pickering, and Richard Norman. 1978. "Some Animals Are More Equal Than Others." *Philosophy* 53 (206): 507–27. doi:10.1017/S0031819100026358.

Freeman, Carrie. 2012. "Fishing For Animal Rights in *The Cove*: A Holistic Approach to Animal Advocacy Documentaries." *Journal for Critical Animal Studies* 10 (1): 104–18.

Freeman, Carrie Packwood, Marc Bekoff, and Sarah M. Bexell. 2011. "Giving Voice to the 'Voiceless.'" *Journalism Studies* 12 (5): 590–607. doi:10.1080/1461670X.2010.540136.

Freud, Sigmund. 1893. "Studies on Hysteria." *The Standard Edition of the Complete Psychological Works of Sigmund Freud*, edited by James Strachey. Vol. 2. London: Hogarth.

Freud, Sigmund. 1900. "The Interpretation of Dreams (Second Part)." *The Standard Edition of the Complete Psychological Works of Sigmund Freud*, edited by James Strachey. Vol. 5. London: Hogarth.

Freud, Sigmund. 1909a. "Analysis of a Phobia in a Five-Year-Old Boy." In "Two Case Histories." *The Standard Edition of the Complete Psychological Works of Sigmund Freud*, edited by James Strachey. Vol. 10. London: Hogarth.

Freud, Sigmund. 1909b. "Notes Upon a Case of Obsessional Neurosis." In "Two Case Histories." *The Standard Edition of the Complete Psychological Works of Sigmund Freud*, edited by James Strachey. Vol. 10. London: Hogarth.

Freud, Sigmund. 1910. "Leonardo Da Vinci and a Memory of His Childhood." In *The Standard Edition of the Complete Psychological Works of Sigmund Freud*. Vol. 11. London: Hogarth Press.

Freud, Sigmund. 1913. "Totem and Taboo." In *The Standard Edition of the Complete Psychological Works of Sigmund Freud*. Vol. 13. London: Hogarth Press.

Freud, Sigmund. 1918. "From the History of an Infantile Neurosis." In *The Standard Edition of the Complete Psychological Works of Sigmund Freud*. Vol. 17. London: Hogarth.

Freud, Sigmund. 1926a. "Inhibitions, Symptoms and Anxiety." In *The Standard Edition of the Complete Psychological Works of Sigmund Freud*. Vol. 20. London: Hogarth Press.

Freud, Sigmund. 1926b. "The Question of Lay Analysis." In *The Standard Edition of the Complete Psychological Works of Sigmund Freud*. London: Hogarth Press.

Freud, Sigmund. 1975. *Letters of Sigmund Freud*, edited by Ernst L. Freud. Translated by Tania and James Stern. New York: Basic Books.

Frey, R. G. 1980. *Interests and Rights: The Case Against Animals*. Oxford: Clarendon Press.

Fudge, Erica. 2002. "A Left-handed Blow: Writing the History of Animals." In *Representing Animals*, edited by Nigel Rothfels, 3–18. Bloomington: Indiana University Press.

Gaard, Greta. 2001. "Tools for a Cross-Cultural Feminist Ethics: Exploring Ethical Contexts and Contents in the Makah Whale Hunt." *Hypatia: A Journal of Feminist Philosophy* 16 (1): 1–26. doi:10.2979/HYP.2001.16.1.1.

Gaarder, Emily. 2011. *Women and the Animal Rights Movement*. New Brunswick, NJ: Rutgers University Press.

Gallegly, Elton. 2009. "Animal Welfare: Cruelty barbaric in itself, and tied to other offenses." *The Hill*, July 27. http://thehill.com/special-reports/animal-welfare-july-2009/55631-animal-welfare-cruelty-barbaric-in-itself-and-tied-to-other-offenses.

Garber, Marjorie B. 1996. *Dog Love*. New York: Simon and Schuster.

Garrett, Aaron. 2007. "Francis Hutcheson and the Origin of Animal Rights." *Journal of the History of Philosophy* 45 (2): 243–65. doi:10.1353/hph.2007.0032.

Gilligan, Carol. 1982. *In a Different Voice: Psychological Theory and Women's Development*. Cambridge, MA: Harvard University Press.

Glaberson, William. 2011. "By Helping a Girl Testify at a Rape Trial, a Dog Ignites a Legal Debate." *New York Times*, August 9, A15.

Glendinning, Simon. 1996. "Heidegger and the Question of Animality." *International Journal of Philosophical Studies* 4 (1): 67–86. doi:10.1080/09672559608570825.

Glick, Megan H. 2013. "Animal Instincts: Race, Criminality, and the Reversal of the 'Human.'" *American Quarterly* 65 (3): 639–59. doi:10.1353/aq.2013.0046.

Greenspan, Stanley I., and Stuart Shanker. 2004. *The First Idea: How Symbols, Language, and Intelligence Evolved from Our Early Primate Ancestors to Modern Humans*. Kindle edition. Cambridge, MA: Da Capo Press.

Grinberg, Emanuella. 2011. "Vegan Diet for Dogs: A Question of Thriving vs Surviving." *CNN*, March 10. http://www.cnn.com/2011/LIVING/03/10/vegan.dog.diet/.

Gross, Terry. 2009. "'Cove' Director Surfaces Deep (And Dark) Secrets." *Fresh Air*. NPR. http://www.npr.org/templates/transcript/transcript.php?storyId=111340010.

Gruen, Lori. 2011. *Ethics and Animals: An Introduction*. Cambridge: Cambridge University Press.

Gruen, Lori. 2015. *Entangled Empathy: An Alternative Ethic for Our Relationships with Animals*. Kindle edition. New York: Lantern Books.

Guenther, Lisa. 2013. *Solitary Confinement: Social Death and Its Afterlives*. Minneapolis: University of Minnesota.

Gutmann, Amy. 1999. "Introduction." In *The Lives of Animals*, edited by J. M. Coetzee and Amy Gutmann, 3–11. Princeton, NJ: Princeton University Press.

Haney-López, Ian. 2003. *Racism on Trial: The Chicano Fight for Justice*. Cambridge, MA: Belknap Press of Harvard University Press.

Harper, A. Breeze. 2011. "Connections: Speciesism, Racism, and Whiteness as the Norm." In *Sister Species: Women, Animals, and Social Justice*, edited by Lisa A. Kemmerer and Carol J. Adams, 72–78. Urbana: University of Illinois Press.

Heidegger, Martin. 1962. *Being and Time*. Translated by John Macquarrie and Edward Robinson. New York: Harper.

Heidegger, Martin. 1968. *What Is Called Thinking?* New York: Harper and Row.

Heidegger, Martin. 1971. *Poetry, Language, Thought*. New York: Harper and Row.

Heidegger, Martin. 1982. "A Dialogue on Language between a Japanese and an Inquirer." In *On the Way to Language*, translated by Peter D. Hertz, 1–54. San Francisco: Harper and Row.

Heidegger, Martin. 1993. *Basic Writings: Second Edition, Revised and Expanded*. San Francisco, CA: HarperSanFrancisco.

Heidegger, Martin. 1995. *The Fundamental Concepts of Metaphysics: World, Finitude, Solitude*. Translated by William McNeill and Nicholas Walker. Bloomington: Indiana University Press.

Heidegger, Martin. 1996. *Being and Time*. Translated by Joan Stambaugh. Albany, NY: SUNY Press.

Heidegger, Martin. 2000. *Introduction to Metaphysics*. New Haven, CT: Yale University Press.

Heidegger, Martin. 2009. *Basic Concepts of Aristotelian Philosophy*. Translated by Robert D. Metcalf and Mark B. Tanzer. Bloomington: Indiana University Press.

Hendricks, Christina, and Kelly Oliver. 1999. "How to Do (Feminist) Things with Words." Introduction to *Language and Liberation: Feminism, Philosophy, and Language*, edited by Christina Hendricks and Kelly Oliver, 1–38. Albany: State University of New York Press.

"High Court Rules Against PETA Holocaust Ad Campaign." 2009. March 26. http://www.thelocal.de/20090326/18261.

Hobbes, Thomas, and Richard Tuck. 1991. *Leviathan*. Cambridge: Cambridge University Press.

Holman, Ryan. 2010. "Animal Rights Group 'Not Queer Enough.'" *Advocate*, February 3. http://www.advocate.com/News/Daily_News/2010/02/03/Animal_Rights_Group_Not_Queer_Enough/.

hooks, bell. 1981. *Ain't I a Woman: Black Women and Feminism*. Boston: South End Press.

"Importing Violence: How Mass Immigration Is Bringing Violent Crime to Your Community." 2005. *Carrying Capacity Network*, May. http://www.carryingcapacity.org/alerts/alert0505.html.

"Israel Renames Unkosher Swine Flu." 2009. *BBC News*, April 27. http://news.bbc.co.uk/2/hi/middle_east/8021301.stm.

Ito, M. 2008. "Seeing Animals, Speaking of Nature: Visual Culture and the Question of the Animal." *Theory, Culture and Society* 25 (4): 119–37. doi:10.1177/0263276408091987.

Kasaona, John. 2010. "Transcript of 'How Poachers Became Caretakers.'" *Ted Talk*, February. https://www.ted.com/talks/john_kasaona_from_poachers_to_caretakers/transcript?language=en.

Kheel, Marti. 2004. "Vegetarianism and Ecofeminism: Toppling Patriarchy with a Fork." In *Food for Thought: The Debate Over Eating Meat*, edited by A. F. Sapontzis, 327–341. Amherst, NY: Prometheus Books.

Kim, Claire Jean. 2014. "The Wonderful, Horrible Life of Michael Vick." In *Ecofeminism: Feminist Intersections with Other Animals and the Earth*, edited by Carol J. Adams and Lori Gruen, 175 190. Kindle edition. New York: Bloomsbury Academic.

Kim, Claire Jean. 2015. *Dangerous Crossings: Race, Species, and Nature in a Multicultural Age*. Kindle edition. New York: Columbia University Press.

Kim, Claire Jean, and Clara Freccero. 2013. "Species/Race/Sex." *American Quarterly* 65 (3).

Lawlor, Leonard. 2007. *This Is Not Sufficient: An Essay on Animality and Human Nature in Derrida*. New York: Columbia University Press.

Laycock, Steven. 1999. "The Animal *as* Animal." In *Animal Others: On Ethics, Ontology, and Animal Life*, edited by H. Peter Steeves, 271–284. Albany, NY: State University of New York Press.

Leahy, Michael P. T. 1991. *Against Liberation: Putting Animals in Perspective*. London: Routledge.

Lee, Jack. 2008. "How Should Animals Be Treated?" *Ethics, Place and Environment* 11 (2): 181–89. doi:10.1080/13668790802252363.

Lee-Lampshire, Wendy. 1995. "Women-animals-machines: A Grammar for a Wittgensteinian Ecofeminism." *Journal of Value Inquiry* 29 (1): 89–101. doi:10.1007/BF01079066.

Liddell, Henry George, and Robert Scott. 1996. *A Greek-English Lexicon*. 9th ed. Oxford: Clarendon Press.

MacKinnon, Catharine. 2004. "Of Mice and Men: A Feminist Fragment on Animal Right." In *Animal Rights: Current Debates and New Directions*, edited by Cass R. Sunstein and Martha C. Nussbaum, 263–276. Oxford: Oxford University Press.

Maclagan, Eric. 1923. "Leonardo in the Consulting Room." *Burlington Magazine for Connoisseurs* 43 (238): 54, 57–58.

Malkin, Michelle. 2009. "Hey, Maybe We'll Finally Get Serious about Borders Now." *Michelle Malkin*, April 25. http://michellemalkin.com/2009/04/25/hey-maybe-well-finally-get-serious-about-borders-now/.

Malphrus, Grant. 2013. "Matter of Agustin ORTEGA-LOPEZ." Cite as 26 I&N Dec. 99 (Bia 2013), and Interim Decision #3777.

Marder, Elissa. 2009. "The Sex of Death and the Maternal Crypt." *Parallax* 15 (1): 5–20. doi:10.1080/13534640802604331.

McKay, Robert. 2010. "Metafiction, Vegetarianism, and the Literary Performance of Animal Ethics in JM Coetzee's *The Lives of Animals*." *Safundi: The Journal of South African and American Studies* 11 (1–2): 67–85.

McKenna, Erin. 1994. "Feminism and Vegetarianism: A Critique of Peter Singer." *Philosophy in the Contemporary World* 1 (3): 28–35.

McKinley, Jesse. 2015. "Judge Orders Stony Brook University to Defend Its Custody of 2 Chimps." *New York Times*, April 21. http://www.nytimes.com/2015/04/22/nyregion/judge-orders-hearing-for-2-chimps-said-to-be-unlawfully-detained.html?_r=0.

Medina, José. 2003. "Identity Trouble: Disidentification and the Problem of Difference." *Philosophy and Social Criticism* 29 (6): 655–80. doi:10.1177/0191453703296002.

Mitchell, W. J. T. 2003. "The Rights of Things." Foreword to *Animal Rites: American Culture, the Discourse of Species, and Posthumanist Theory*, by Cary Wolfe, ix–xiv. Chicago: University of Chicago Press.

Molnar, Michael. 1996. "Of Dogs and Doggerel." *American Imago* 53 (3): 269–80. doi:10.1353/aim.1996.0010.

Monk, Ray. 1990. *Ludwig Wittgenstein: The Duty of Genius*. New York: Free Press.

Mulhall, Stephen. 2009. *The Wounded Animal: JM Coetzee and the Difficulty of Reality in Literature and Philosophy*. Princeton, NJ: Princeton University Press.

Nagel, Thomas. 1974. "What Is It Like to Be a Bat?" *Philosophical Review* 83 (4): 435–50. doi:10.2307/2183914.

Napolitano, Joseph D. 2010. "'Mr Melancholy and Mr Magpie': The Lives of Animals in JM Coetzee's *Diary of a Bad Year*." *Safundi: The Journal of South African and American Studies* 11 (1–2): 49–66.

Oliver, Kelly. 2009. *Animal Lessons: How They Teach Us to Be Human*. New York: Columbia University Press.

Oliver, Kelly. 2012. "Ambivalence Toward Animals and the Moral Community." *Hypatia* 27 (3), 493–498. doi:10.1111/j.1527-2001.2012.01291_2.x.

O'Neill-Stephens, Ellen. 2010. "Legal Support for the Use of a Facility Dog to Assist Testifying Crime Victims and Witnesses: A Review of Evidence Rule, Case Law, State Statutes, and Legislation." http://courthousedogs.org/pdf/CourthouseDogs-LegalSupport.pdf April 21, 2015.

Ornstein, Anna. 1993. "Little Hans: His Phobia and His Oedipus Complex." In *Freud's Case Studies: Self-Psychological Perspectives*, edited by Barry Magid, 87–106. Hillsdale, NJ: Analytic Press.

Otto, Stephan. 2011. "Peacocks, Dogfighting & Sexual Assault—A Legislative Update—Animal Legal Defense Fund." *Animal Legal Defense Fund*, May 9. http://aldf.org/blog/peacocks-dogfighting-sexual-assault-a-legislative-update/.

Otto, Stephan, ed. 2011. "Animal Protection Laws of the United States of America and Canada." *Animal Legal Defense Fund*. http://aldf.org/resources/advocating-for-animals/animal-protection-laws-of-the-united-states-of-america-and-canada/.

Parikka, Jussi. 2010. *Insect Media: An Archaeology of Animals and Technology*. Minneapolis: University of Minnesota Press.

Peacock, Laurel. 2009. "Animots and the Alphabête in the Poetry of Francis Ponge." *Australian Humanities Review* 47: 89–97.

Pellegrini, Ann. 2009. "The Dogs of War and the Dogs at Home: Thresholds of Loss." *American Imago* 66 (2): 231–51. doi:10.1353/aim.0.0048.

Peterson, Christopher. 2010. "The Aping Apes of Poe and Wright: Race, Animality, and Mimicry in 'The Murders in the Rue Morgue' and *Native Son*." *New Literary History* 41 (1): 151–71. doi:10.1353/nlh.0.0141.

Phillips, Dana. 2010. "Weeping Elephants, Sensitive Men." *Safundi: The Journal of South African and American Studies* 11 (1–2): 19–47.

Pleasants, Nigel. 2006. "Nonsense on Stilts? Wittgenstein, Ethics, and the Lives of Animals." *Inquiry* 49 (4): 314–36. doi:10.1080/00201740600831364.

Psihoyos, Louie, and Mark Monroe. 2009. *The Cove*. DVD.

Raber, Karen. 2008. "How to Do Things with Animals: Thoughts On/with the Early Modern Cat." In *Early Modern Ecostudies: From the Florentine Codex to Shakespeare*, edited by Ivo Kamps, Karen L. Raber, and Thomas Hallock, 93–113. New York: Palgrave Macmillan.

Regan, Tom. 1985. "The Case for Animal Rights." In *In Defense of Animals*, edited by Peter Singer, 13–26. New York: Blackwell.

Regan, Tom. 2003. *Animal Rights, Human Wrongs: An Introduction to Moral Philosophy*. Lanham, MD: Rowman & Littlefield.

Regan, Tom. 2004. *The Case for Animal Rights: Updated with a New Preface*. Berkeley: University of California Press.

Roberts, Mark S. 2008. *The Mark of the Beast: Animality and Human Oppression*. West Lafayette, IN: Purdue University Press.

Rohman, Carrie. 2009. *Stalking the Subject: Modernism and the Animal*. New York: Columbia University Press.

Rothfels, Nigel. 2002. *Representing Animals*. Bloomington: Indiana University Press.

Rousseau, Jean-Jacques, and Johann Gottfried Herder. 1986. *Essay on the Origin of Language*. Chicago: University of Chicago Press.

Russell, Bertrand. 1951. "Obituary: Ludwig Wittgenstein." *Mind* 60 (239): 297–98.

Sapontzis, S. F. 2004. *Food for Thought: The Debate over Eating Meat*. Amherst, NY: Prometheus Books.

Schecter, Anna. 2014. "Tyson Foods Changes Pig Care Policies after NBC Shows Undercover Video." *NBC News*, January 10.
http://investigations.nbcnews.com/_news/2014/01/10/22245308-tyson-foods-changes-pig-care-policies-after-nbc-shows-undercover-video

Schlosser, Eric. 2001. "The Chain Never Stops." *Mother Jones*. http://www.motherjones.com/politics/2001/07/dangerous-meatpacking-jobs-eric-schlosser.

Schneider, Elizabeth. 1990. "The Dialectic of Rights and Politics: Perspectives from the Women's Movement." In *Women, the State, and Welfare*, edited by Linda Gordon, 226–249. Madison, WI: University of Wisconsin Press.

Schwartzman, Lisa. 1999. "Liberal Rights Theory and Social Inequality: A Feminist Critique." *Hypatia: A Journal of Feminist Philosophy* 14 (2): 26–47. doi:10.2979/HYP.1999.14.2.26.

Seneca. 2013. *Of Clemency (Annotated) (Dialogues of Seneca Book 6)*. Translated by Aubrey Stewart. Kindle edition. Amazon Digital Services.

Seshadri, Kalpana. 2012. *HumAnimal: Race, Law, Language*. Minneapolis: University of Minnesota Press.

Seshadri-Crooks, Kalpana. 2003. "Being Human: Bestiality, Anthropophagy, and Law." *Umbr(a)* 1: 97–116.

Shammas, Brittany. 2015. "Children Practice Reading Aloud to Dogs in 'Tail Waggin' Tutor' Program." *SunSentinel*. http://www.sun-sentinel.com/local/palm-beach/fl-boca-reading-therapy-dogs-20150417-story.html.

Shapiro, Paul. 2006. "Moral Agency in Other Animals." *Theoretical Medicine and Bioethics* 27 (4): 357–73. doi:10.1007/s11017-006-9010-0.

Singer, Peter. 2009. *Animal Liberation: The Definitive Classic of the Animal Movement*. New York: Harper Perennial.

"Society of Professional Journalists Code of Ethics." 2014. *SPJ*. http://www.spj.org/ethicscode.asp.

Solomon, Olga. 2010. "What a Dog Can Do: Children with Autism and Therapy Dogs in Social Interaction." *Ethos* 38 (1): 143–66.

Sorabji, Richard. 1993. *Animal Minds and Human Morals: The Origins of the Western Debate*. Ithaca, NY: Cornell University Press.

Steiner, Gary. 2005. *Anthropomorphism and Its Discontents: The Moral Status of Animals in the History of Western Philosophy*. Pittsburgh: University of Pittsburgh Press.

Sunstein, Cass. 2004. "Can Animals Sue?" In *Animal Rights: Current Debates and New Directions*, edited by Cass R. Sunstein and Martha C. Nussbaum, 251–262. Oxford: Oxford University Press.

Sunstein, Cass, and Jeff Leslie. 2010. "Animal Rights without Controversy." *Law and Contemporary Problems* 70: 117–38.

Sunstein, Cass R., and Martha C. Nussbaum. 2004. *Animal Rights: Current Debates and New Directions*. Oxford: Oxford University Press.

Tarver, Erin C. 2013. "The Dangerous Individual('s) Dog: Race, Criminality and the 'Pit Bull.'" *Culture, Theory and Critique*: 1–13. doi:10.1080/14735784.2013.847379.

Taylor, Paul W. 1987. "Inherent Value and Moral Rights." *Monist* 70 (1): 15–30. doi:10.5840/monist198770113.

Taylor, Sunaura. 2014. "Interdependent Animals: A Feminist Disability Ethic-of-Care." In *Ecofeminism: Feminist Intersections with Other Animals and the Earth*, edited by Carol J. Adams and Lori Gruen, 109–126. Kindle edition. New York: Bloomsbury Academic.

Vizcarrondo, Sara. 2009. "Dolphin Confidential: Docu-Thriller 'The Cove' Exposes Slaughter." *International Documentary Association*. July. http://www.documentary.org/feature/dolphin-confidential-docu-thriller-cove-exposes-slaughter.

Waal, F. B. M. de. 2006. *Primates and Philosophers: How Morality Evolved*, edited by Stephen Macedo and Josiah Ober. Kindle edition. Princeton, NJ: Princeton University Press.

Warren, Mary Anne. 1997. *Moral Status: Obligations to Persons and Other Living Things*. Oxford: Clarendon Press.

Weil, K. 2010. "A Report on the Animal Turn." *Differences* 21 (2): 1–23. doi:10.1215/10407391-2010-001.

Willett, Cynthia. 2014. *Interspecies Ethics*. Kindle edition. New York: Columbia University Press.

Williams, Patricia J. 1991. *The Alchemy of Race and Rights*. Cambridge, MA: Harvard University Press.

Winkler, Rafael. 2007. "Heidegger and the Question of Man's Poverty in World." *International Journal of Philosophical Studies* 15 (4): 521–39.

Wise, Steven M. 2002. *Drawing the Line: Science and the Case for Animal Rights*. Cambridge, MA: Perseus Books.

Wittgenstein, Ludwig. 1967. *Zettel*. Translated by G. E. M. Anscombe. Berkeley: University of California Press.

Wittgenstein, Ludwig. 2009. *Philosophical Investigations* (4th edition), edited by P. M. S. Hacker and Joachim Schulte. Malden, MA: Wiley-Blackwell.

Wolfe, Cary. 2003. *Animal Rites: American Culture, the Discourse of Species, and Posthumanist Theory*. Chicago: University of Chicago Press.

Wolfe, Cary. 2008. "Flesh and Finitude: Thinking Animals in (Post)Humanist Philosophy." *SubStance* 37 (3): 8–36. doi:10.1353/sub.0.0015.

Wolfe, Cary. 2010. "Before the Law: Animals in a Biopolitical Context." *Law, Culture and the Humanities* 6 (1): 8–23. doi:10.1177/1743872109348986.

Wolfe, Cary. 2013. *Before the Law: Humans and Other Animals in a Biopolitical Frame*. Kindle edition. Chicago: University of Chicago Press.

Wyckoff, Jason. 2015. "The Problem of Speaking for Animals." In *Animal Ethics and Philosophy: Questioning the Orthodoxy*, edited by Elisa Aaltola and John Hadley, 117–132. Kindle edition. London: Rowman and Littlefield International.

Zimmerman, Michael. 1983. "Toward a Heideggerian Ethos For Radical Environmentalism." *Environmental Ethics* 5: 99–132.

# Index

www.ingramcontent.com/pod-product-compliance
Lightning Source LLC
LaVergne TN
LVHW101448200926
848483LV00001B/95